FROM THE GROUND UP

Re-materialising Cultural Geography

Dr Mark Boyle, Department of Geography, University of Strathclyde, UK,
Professor Donald Mitchell, Maxwell School, Syracuse University, USA and
Dr David Pinder, Queen Mary University of London, UK

Nearly 25 years have elapsed since Peter Jackson's seminal call to integrate cultural geography back into the heart of social geography. During this time, a wealth of research has been published which has improved our understanding of how culture both plays a part in, and in turn, is shaped by social relations based on class, gender, race, ethnicity, nationality, disability, age, sexuality and so on. In spite of the achievements of this mountain of scholarship, the task of grounding culture in its proper social contexts remains in its infancy. This series therefore seeks to promote the continued significance of exploring the dialectical relations which exist between culture, social relations and space and place. Its overall aim is to make a contribution to the consolidation, development and promotion of the ongoing project of re-materialising cultural geography.

Also in the series

Sanctuaries of the City
Lessons from Tokyo
Anni Greve
ISBN 978 0 7546 7764 2

Cities and Fascination
Beyond the Surplus of Meaning
Edited by Heiko Schmid, Wolf-Dietrich Sahr and John Urry
ISBN 978 1 4094 1853 5

Swinging City
A Cultural Geography of London 1950–1974
Simon Rycroft
ISBN 978 0 7546 4830 7

Remembering, Forgetting and City Builders
Edited by Tovi Fenster and Haim Yacobi
ISBN 978 1 4094 0667 9

Doing Family Photography
The Domestic, The Public and The Politics of Sentiment
Gillian Rose
ISBN 978 0 7546 7732 1

From the Ground Up
Community Gardens in New York City and the Politics of Spatial Transformation

EFRAT EIZENBERG
Technion, Israel Insitute of Technology

Routledge
Taylor & Francis Group

LONDON AND NEW YORK

First published 2013 by Ashgate Publishing

2 Park Square, Milton Park, Abingdon, Oxon OX14 4RN
711 Third Avenue, New York, NY 10017, USA

Routledge is an imprint of the Taylor & Francis Group, an informa business

First issued in paperback 2016

British Library Cataloguing in Publication Data
Eizenberg, Efrat.
 From the ground up : community gardens in New York City and
 the politics of spatial transformation. --
 (Re-materialising cultural geography)
 1. Community gardens--Social aspects--New York (State)--
 New York. 2. Community gardens--Political aspects--New
 York (State)--New York. 3. Community development--New
 York (State)--New York. 4. Social participation--New York
 (State)--New York. 5. City dwellers--Political activity--
 New York (State)--New York.
 I. Title II. Series
 307.'216'097471-dc23

Library of Congress Cataloging-in-Publication Data
Eizenberg, Efrat.
 From the ground up : community gardens in New York City and the politics of spatial
transformation / by Efrat Eizenberg.
   p. cm.
 Includes bibliographical references and index.
 ISBN 978-1-4094-2909-8 (hardback) 1. Urban gardens--New York (State)--New York. 2. Urban
gardens--Social aspects--New York (State)--New York. 3. Gardens--New York (State)--New York.
4. Public spaces--Political aspects--New York (State)--New York. I. Title.
 SB454.3.S63E37 2012
 635.09747'1--dc23

                                                                                    2012021241

ISBN 978-1-4094-2909-8 (hbk)
ISBN 978-1-138-26145-7 (pbk)

Contents

To Roya

List of Figures, Maps and Table

Figures

Maps

Table

Acknowledgments

This research and book were made possible because of the many community gardeners of New York City and green organizations that invited me into their world and shared with me their knowledge, activities, and vision. Without their kindness, smartness, and enthusiasm this research would not have been the enjoyable experience it was. Most notably, I would like to thank the amazing people (too many to mention by name) at the New York City Community Gardens Coalition, the Green Guerillas, and the gardeners at La Plaza Cultural Community Garden for their friendship and for showing me the colors and beauty within the grayness, literally and figuratively. I hope to take with me wherever I go everything I learned from them about our relations to the environments we inhabit, about the political quintessence of human beings and the important things in life.

In this decade-long process I was very lucky to have four wonderful scholars and friends from The Graduate Center, CUNY who worked and fertilized the soil on which ideas and words grew and flourished: Prof. Susan Saegert with whom I worked on several research projects, and who taught me the inside out of doing research with communities, Professor Cindi Katz whose theoretical contributions and careful editing were invaluable. Professor Stanley Aronowitz who introduced me to a new world of theory that became the core of my work and who identified the potential of my work from the first paper I submitted to him, and Professor Roger Hart who supported me from the first moment and guided me with endless courtesy and care. His absolute belief in me was the engine that propelled me forward. Each of them was a mentor to me and each nurtured me in a unique way. I believe and hope that their voices are echoed through my work and that their scholarship will keep guiding mine. I am also thankful to Professor Tovi Fenster from Tel Aviv University for providing the time, support and encouragement that help these ideas and words to reap into a complete manuscript.

Many other people supported my work with ideas, friendship or both. Especially I would like to thank Elad Eizenberg for producing the diagrams and other visual materials, to Carol Fisher for contributing the front cover image and to Yossef Bodovski for his enormous help with producing the maps.

Finally, my deepest gratitude to my dear parents and brother Nilli, Hezi and Elad Eizenberg for their endless love and to my family and my good friends for their love and support and to Eran Fisher for walking with me through its difficult and pleasant segments. They all allowed this long journey to take place, encouraged me, and found ways to make it not only possible but also meaningful and enjoyable. And Roya, I thank you for being part of my life and for the joy and beauty that you spread around. I hope that your future habitat would maintain many opportunities for different urban experiences as was envisioned in community gardens.

Introduction

Aloof, withdrawn into themselves, over-intellectualized, maintaining impersonal relationships, dependent on strangers to sustain their everyday conduct, their efforts and interests are narrowed down to the economic system, embodying an identity of producers/consumers. In this way Zimmel characterized urbanites over a century ago—urban residents who struggle to survive in the jungle of overly stimulating, rationalized, and impersonal environment. This is what urban life seemed to offer through remolding the personality and social relations of its inhabitants. What are the chances of urban residents leading lives different from what they were doomed to by this environment? Can they overcome what it prescribes them to be? Are there other opportunities hidden in the city of a different urban experience?

These are the questions this book wishes to offer a possible answer. Unlike rural homogeneity, every aspect of the city is diversified. Therefore it encompasses cracks or openings in which alternative urban life can be carried out. Those cracks are usually not out there in the open but are presented to us very shyly and humbly. The urban space is so rigidly constructed to maintain its essential functions as defined by modernity—work, consumption, education, housing—that neither a bird's-eye view of the city nor strolling the streets may divulge them. Rather, they may be discovered and unearthed only through search. Being both hidden and dispersed, the question of whether they can and will amount to something that eventually transforms urban practice and experience hovers over them.

Nevertheless, recent events suggest that we might have witnessed and experienced a sort of a loud rupture rather than hidden cracks. When this book was already written and almost sealed, reality introduced us (again) with new materials for considering these questions. The 2011 rupture of urban awakening in many cities in the Middle East, Europe, and the US for the first time in such a fury since the uprising of 1968, and in many of the sites for the first time in modern history. Massive numbers of people took over the streets, not only for temporal acts of protesting but also in order to appropriate highly privatized—in spirit and practice—public spaces. In them they worked together to produce a new type of urbanite. They established a strong solidarity that is based on, to summarize in few words, their need and aspiration to become political and to make new politics.

> For many years we have been operating in the city quietly, modestly. But now we need to show our muscles, show our strength, and we can do so through showing our numbers and presenting a united front. We have a very important role in the city, we are the stewards of land and we have to be more proactive about the way we would like this city—our city—to treat its space. We need to make sure to protect our special places where we cultivate communities, where people

can be heard and can affect their own life. And then we have to think of future generations and make sure their city will still protect those spaces. Therefore we need to push for policy that protects [keeps] those spaces and enable us to create more of them. Moreover, it is not only about our city, it is about our state and our nation and earth—we need to learn from others and collaborate with others.

These words were not spoken in one of those recent urban protests; rather, they opened the Second New York City Community Gardeners' Annual Forum in 2006. They are part of an inspirational talk, directed at community gardeners from across the city and to representatives of municipal and supporting organizations who arrived on a rainy Saturday morning to discuss and resolve together the burning issues that gardeners in New York City encounter. This event took place in relatively peaceful times for community gardeners. The settlement that preserves many gardens from destruction was signed with the city in 2002. Back in 2006 they still had four more years until the agreement expires and a new agreement for community gardens in the city has to be reached. These were interim times when the mist of celebration and victory was still felt in the air, but dark clouds were already on the horizon. These nearing clouds were the main reason for bringing gardeners together in a forum; clouds that carried the uncertain future of gardens reminded the organizers that there were still many actions to be taken and aspirations not yet fulfilled.

The story of the community gardens movement in New York City, of community gardeners and their continuous struggle over the city, is a story of reimagining the city with the politics of hope. Central to this story is the firm "belief in the actualities of change that can arise from the unexpected reaction to the vagaries of urban life, the novel organizations that can arise, and, in general, the invention of new spaces of the political."[1] And mostly it is the story of urban residents whose spatial practices and "crystallized visions" of the city bring about a change in their lives, neighborhoods, social relations, and the urban discourse at large. It is the story of the community gardeners of New York City.

Through an interdisciplinary approach, this book analyzes the struggle of urban residents to constitute meaningful and supportive space for themselves and to constitute themselves as meaningful actors on the urban scene. A central goal of this endeavor is to portray the politicizing power of the production of space. Given the political nature of the production of space, it shows the dialectical dynamic in which gardeners constitute the space and are also being constituted as political actors.

The process of the production of space as manifested in community gardens in New York City is taking place at three interrelated levels: individual, collective, and institutional—a tri-layered phenomenon. Respectively the book is divided into three parts, each discussing one of these levels of analysis. Each part describes how everyday practices, social interactions, interactions with the environment,

---

1   Ash, A. and N. Thrift. 2002. *Cities: Reimagining the Urban.* Cambridge: Polity Press, p. 4.

and generation of knowledge with regard to space and the community complicate the awareness of residents, develop their political consciousness, and constitute them as important social actors in the urban scene.

With this multi-level analysis the book registers a wide range of experiences and draws the connection between levels that are usually treated separately in social sciences. Each level is shown to be dependent on the others for its continuation; each informs the other and together they establish an autopoietic force.

The integration of these three levels allows the work to go beyond the traditional understandings of 'politics' or 'political' and register a spectrum of possible connections between spatial practices and political consciousness. Understanding the political potency of spatial practices goes from the individual level to the collective and to the institution of community gardens and back, hence, these three levels are modular and interdependent. The political prowess of gardeners in urban struggles is incomprehensible without connecting it to the individual level, where individuals form an affective connection to space and a sense of control over it. Likewise, individual practices are driven by the energy of the group and its shared spatial practices.

The analytical distinction between the three levels also serves a temporal argument. This work shows a continuum of development whereby political consciousness is developed and in turn informs practice. Along this continuum urban residents become different kinds of social actors with newly acquired skills and capacities which are more informed and reflexive, and ultimately more politicized. They become what I term "organic residents." These urbanites constantly engage with their environment, find their own ways to make it a supportive environment for their collective needs, and produce it in their own image. In other words, the move from the individual, through the collective, and on to the institutional levels is developmental and modular; it is interdependent and dialectical; and it is political in essence.

While expanding on existing literature on community gardens and the various organizations that support and oversee them, this book assembles a contemporary ethnography of community gardens and captures the potential of space to generate a cycle of the development of political consciousness among urban residents that reproduces them as a different type of urbanite, with alternative urban experience and with more fertile grounds for different forms of politics.

Urban Space and Capitalist Strategy

Examining spatial practices like those exhibited in community gardens is set within a framework of social analysis commonly referred to as the "spatial turn," which asserts the centrality of space in understanding social arrangements. With the spatial turn, space has come to be understood as an arena that does not simply reflect but rather constitutes, produces, and reproduces social arrangements. Space and control over space are seen as principal coordinates in shaping society and as a prism through which we can interpret society and reconstruct its social history. The history of spaces, as Foucault taught us, "would be at the same time the history of powers." The intricate and varied factors involved in shaping contemporary space elongate this history "from the great strategies of geopolitics to the little tactics of the habitat."[2] As a social product, the urban environment is constituted through conflicting social interests and values.[3]

Urban space and its transformation are at the core of this work. Urban space, understood here within Marxist traditions of dialectical materialism as developed by Henri Lefebvre, at one and the same time reproduces and reifies social relations. Thus, socially constructed space is never politically neutral; it is always embedded in struggle and conflict; it is always in flux; but it therefore also holds the potential for social transformation. The history of urban transformations has been inextricably linked with capitalism: capitalist development depends on urbanization. Within the process of urbanization, space has become a dominant means for using, producing, and controlling economic surplus. Urban capitalist space geographically concentrates labor and capital, merchants and infrastructure,[4] while neoliberalism—the dominant political-economic strategy of late capitalism—"uses space as its privileged instrument."[5] Moreover, space itself is perceived and acted upon as a commodity, and its increasing commodification influences virtually every cultural, economic, and political institution that operates on the urban scene.[6] In such constellation, the logic of investment and capital circulation shapes the urban political and social structure as well as explains the spatial organization of everyday life.[7]

Urban transformations, the other core element of this work, evolve and come about out of actions of and constant tensions between the forces operating on space. On the one hand, the hegemonic force strives to fully appropriate space into the system of capital accumulation by means of commodification, commercialization, and enclosure, and to better suit it for capitalist progress. For example, gentrification has been generalized since the 1980s as a global strategy of urban expansion and reconstruction. It become the main framework to configure and implement urban

---

2   Foucault (1980) in Soja, E. 1989. *Postmodern Geographies: The Reassertion of Space in Critical Social Theory*. New York: Verso, p. 21.

3   Harvey, D. 1989. *The Urban Experience*. Baltimore: Johns Hopkins University Press; Lefebvre, H. 2003. *The Urban Revolution*. Minneapolis: University of Minnesota Press.

4   Harvey 1989.

5   Benner, N. and N. Theodor. 2002. "Cities and the Geographies of 'Actually Existing Neoliberalism." *Antipode* 34(3): 349–79, p. 343.

6   Harvey 1989; Lefebvre, H. 1991. *The Production of Space*. Oxford: Blackwell; Logan, J. and H. Molotch. 1987. *Urban Fortunes: The Political Economy of Place*. Berkeley: University of California Press.

7   Tajbakhsh, K. 2001. *The Promise of the City: Space, Identity, and Politics in the Contemporary Social Thought*. Berkeley: University of California Press.

policy;[8] a means of embedding the logistics, threads, and assumptions of capital accumulation more deeply than ever in the urban landscape.

The hegemonic production of space is guided by an instrumental rationality according to which space is reproduced to allow for the maximization of its exchange-value and control over space for the ultimate purpose of capital accumulation. To this end, modern space is delineated and divided to accommodate designated activities such as commerce, industry, and residence. It is homogenized to accommodate the smooth conduct of one unitary set of principles and practices, that of capitalism. As a result, space is increasingly and excessively fragmented and compartmentalized, with each segment endowed with a specific function. This transformation of space is made possible mainly through the process of privatization—the transfer of space from the public domain to the market realm.[9]

The tensions that permeate the production of space keep space in flux, in negotiation, and in constant transformation. Responding to challenges and contestations, the process of implementing neoliberal principles in space is constantly changing its methods, creating a "shifting landscape of experimentation."[10] While before the 1980s the neoliberal strategy was characterized by efforts to deregulate the market and minimize state power ("roll-back neoliberalism"), since the 1980s a new strategy, "roll-out neoliberalism," has been mobilized in political management and interventions in order to overcome frictions and recessions and to further institutionalize the neoliberal project.[11] Hence, while neoliberal space is unified by meaning, purpose, and usage, its mechanisms of production are constantly being adjusted in a trial and error fashion.

The tensions that are intrinsic to space that result in spatial transformations are also related to the possible "reaction to the vagaries of urban life."[12] In the uneven flow of capital there will always be places in which capital is not invested, spaces that are marginalized by capitalism. At these margins, spaces can be produced or maintained as *other*, differential spaces that do not conform to hegemonic space. The concurrent existence of hegemonic space and differential space encapsulates the dialectical nature of space. Space envelops dialectical trends of power and resistance, hegemonic space and socially produced contested space, homogeneous space and differential space. This potential "opening" for social change *through space* that is embedded within the process of *the production of space*[13] is at the heart of this book.

---

8   Smith, N. 2002. "New Globalism, New Urbanism: Gentrification as Global Urban Strategy." *Antipode* 34(3): 427–50.

9   Lefebvre 1991.

10   Peck, J. and A. Tickel. 2002. "Neoliberalizing Space." *Antipode* 34(3): 380–404, p. 396.

11   Ibid.

12   Ash and Thrift 2002, p. 4.

13   Harvey, D. 2000. *Spaces of Hope*. Berkeley: University of California Press; Lefebvre 1991.

Thus, for example, the process of neoliberal globalization has turned cities against each other in global (that is, extra-local) competition, and has harmed many locales.[14] Cities compete to attract large investments, to develop their space as touristic sites, and to invest in infrastructure better suited for the global economy, mainly addressing the needs of the FIRE industries (that is, finance, insurance, and real estate). As a result of producing space to satisfy its extra-local users (the tourist, the mobile elite), the locale—local needs, local services, and local investment—has been neglected.[15]

In the 1970s, as urban centers went through intensive restructuring and some major urban centers were established as global cities,[16] some of their locales—neglected and disinvested—became sites for "other" spaces, sites where alternatives could flourish. And they did. Community gardens in New York City were produced on the most neglected locales as sites that contrast and contest their surroundings. They sprang up in the places that capital had left behind undeveloped. But with changing neoliberal strategies these same neglected locales became the targets for neoliberal experimentation with new local entrepreneurial culture.[17] Differential spaces constituted as community gardens interfered with those experimentations with "new localism" and the window of opportunity for community gardens to thrive was beginning to close. The existence of community gardens side by side with existant neoliberal spaces constitutes the locale as a contested arena of opposites and ambiguities, and as paradigmatic sites for the examination of struggles over space and the spatially embedded potentialities of social change.

Differential Space

If social relations produce space and are reproduced by it, they can be altered by actions that transform space. That is, social change can come about through forces demanding and reclaiming space and changing the conditions of its production.

---

14   Hackworth, J. 2007. *The Neoliberal City: Governance, Ideology, and Development in American Urbanism*. Ithaca, NY: Cornell University Press; Site, W. 2003. *Remaking New York: Primitive Globalization and the Politics of Urban Community*. Minneapolis: University of Minnesota Press.

15   Bauman, Z. 1998. *Globalization: The Human Consequences*. New York: Columbia University Press; Sassen, S. 1998. *Globalization and its Discontents*. New York: The New Press.

16   Sassen 1998.

17   Benner and Theodor 2002. See also examples in Davilla, A. 2004. *Barrio Dreams: Puerto Ricans, Latinos, and the Neoliberal City*. Berkeley: University of California Press; Lloyd, R. 2005. *Neo-Bohemia: Art and Commerce in the Postindustrial City*. New York: Routledge; Mele, C. 2000. *Selling the Lower East Side: Culture, Real Estate, and Resistance in New York City*. Minneapolis: University of Minnesota Press.

Space is the locus of social change; the promise of liberation from repression and exploitation can spring and materialize through the production of *differential space*.[18]

While the dominant production of space suppresses the concrete qualities of space, differential space celebrates exactly these qualities. It allows for "bodily and experiential particularity" to take place, and promotes "the nonnegotiable 'right to difference'."[19] A differential space is an actual place that exists as "counter-space"; other spaces of society may be represented in it, but they are at the same time contested and inverted. It represents a negation of the dominant space, that which is being produced by the dominant social relations and mode of production. Space's dialectical trends are manifested in the existence of differential spaces within the dominant space. These trends are commonly termed "isotopy" and "heterotopias." Heterotopias (plural), places that represent multiplicity, are marked by their difference from the dominant environment of the isotopy (always singular). The difference is a relational not an objective one and "can extend from a highly marked contrast all the way to conflict."[20] For example, archetypical heterotopias in the contemporary isotopy are those which negate commercial exchange.[21] Similarly, community gardens represent a negation of the dominant logic and practice in their emphasis on the use value of space rather than its exchange value. Differential spaces, one of the materializations of heterotopias, are constituted as spaces of alternate ordering in which the social world is organized differently and in which life is experienced differently; as such they disrupt the homogeneity of society.[22]

Unlike the notion of heterotopias as spaces that spark off almost randomly in relation to time and space,[23] differential spaces as manifestations of heterotopias are understood as a critical register that is embedded in a sense of political and historical deviance from social norms. Thus differential spaces do not evolve uncontrollably or disorderly as a result of a random change of power. Rather, differential space is intentional, hence political; while capitalism strives to enclose space, differential space is an effort to open and reopen space for change. Appropriation of space and its alternative production amount to oppositional practices.[24]

By unearthing the *particular* practices and processes entailed by the production of space of community gardens, this analysis of community gardens offers three dimensions of understanding. First, it unpacks and reveals the relations of production of the space of community gardens in New York City. The analysis of the space of community gardens is being used here as a lens into macro processes

---

18   Lefebvre 2003.

19   Merrifield, A. 2005. *Henri Lefebvre: A Critical Introduction*. New York: Routledge, p. 113.

20   Lefebvre 2003, p. 38.

21   Examples of heterotopias are a commune or a Kibbutz within the capitalist state (uncommodified space within a generally commodified one).

22   Harvey 2000.

23   Foucault, M. 1986. "Of Other Spaces." *Diacritics* 16(1): 22–7.

24   Lefebvre 2003.

and power relationships acting in the urban environment. Second, it presents the space of community gardens as a differential space that takes the concrete form of the (modern) *commons*. It shows how such differential space is being produced, maintained and in turn affects the surrounding environment. Third, it supplements the analysis of urban political economy with an interdisciplinary analysis of political practices. It analyzes the interaction between spatial practices and political practices which take place in and around community gardens. It shows how individual and collective spatial practices of "survival" are in effect being translated into politically charged consciousness and actions.

Moving from individual practices of survival towards collective political practices is supported by the model of *resistance* developed by Katz (2004). Katz deconstructs the notion of resistance into three modular moments. She delineates practices of opposition and transformation, which are generally termed "resistance," into three different modes of action and consciousness:[25] *resilience, reworking*, and *resistance* (the three Rs of resistance). These modes of practice are manifestations of the reactions of oppressed individuals, communities, and people to the social, economic, and political conditions of late modernity and neoliberal regimes. Briefly stated (and further explained later on), *resilience* encompasses practices of survival, the capacity of people to find refuge from oppressive life conditions. *Reworking* refers to practices that alter life conditions. Reworking practices involve recognition of the problems and are aimed at solving them within the same register in which they are experienced. Both resilience and reworking practices, according to Katz, have the potential to develop into oppositional consciousness and practices of resistance. Practices of resistance involve oppositional consciousness and are aimed at solutions on a scale broader than specific local problems.

The three levels of production of space in community gardens are explained with the model of the three Rs of resistance. Thus each level of the process of social production of space of community gardens—from the individual through the collective and to the institution—is shown to be an expression of resilience, reworking, and resistance practices. Katz's model of resistance also helps develop the argument on the centrality of space as a coordinate that propels such practices of resilience, reworking, and resistance. Katz's model features play behavior and playfulness as its central coordinate and identifies the mimetic faculty as the "means to spark consciousness and provoke an alternative, oppositional, and even revolutionary imagination that can see in the same, something different."[26] This

---

25   Katz's model of resistance ties together action and consciousness from a developmental approach and therefore fits my analysis better than other delineations of resistance. Katz, C. 2004. *Growing Up Global: Economic Restructuring and Children's Everyday Lives*. Minneapolis: University of Minnesota Press. See Bayat, A. 2000. "From 'Dangerous Classes' to 'Quiet Rebels': Politics of the Urban Subaltern in the Global South." *International Sociology* 15(3): 533–57 for a review of other delineations of resistance.

26   Katz 2004, p. 257.

faculty enables the development of consciousness regarding the actual world and the imagination of a different one. As such it helps to denaturalize what is "natural" or "normal" and find cracks in what is being presented as "inevitable." In this book I borrow the structure of the model but replace the mimetic faculty with spatial practices as its central coordinate. That is, I wish to show the potential power of spatial practices to recalibrate the imagination, disclose new possibilities of being in the world, and spark new energies that can amount to resilience, reworking, and resistance practices. It might even, as Katz suggests, lead to a revolutionary imagination or a revolution *per se*.

Space and Social Transformation

Community gardens are seen as caught within and reacting to two significant trends of the modern production of space. The first trend is the "abstractification" of space which narrows down its various concrete qualities into abstract value. This trend was for example identified as the *property regime*—the organization and quantification of space as a commodity.[27] The second and interrelated trend is the detachment of users from their space. As space is seen mainly for its exchange value, its use value (and users) is being disintegrated. These two trends are elaborated on throughout the deconstruction of space to its elements, and the framework of the "right to the city."[28]

From Abstract Space to Political Space

Modern space is increasingly abstracted. In the same way that labor was abstracted and hence dissociated from the worker in order for capitalism to arise, so is space being fetishized and the relations of its production masked. Space is perceived as a unitary, unproblematic entity that is "confined within the … frameworks of the dominant relations and mode of production."[29] It is therefore necessary to "unpack" space and reveal the social relations that produced it as well as the social relations it produces. Moreover, "within abstract space are subtle ideological and political machinations, which maintain a perpetual dialogue between its space and users, promoting compliance and 'nonaggression' pacts."[30] The unpacking of space is therefore not only an intellectual task but also a required political task.

Following the dialectical model of Marx, Lefebvre denaturalizes space by unfolding space into its elements. Space envelops a triad of interlocked elements:

---

27 Blomley, N. 2004. *Unsettling the City: Urban Land and the Politics of Property.* New York: Routledge.

28 Lefebvre 1991; Mitchell, D. 2003. *The Right to the City: Social Justice and the Fight for Public Space.* New York: The Guilford Press.

29 Lefebvre 1991, p. 90.

30 Merrifield 2005, p. 112.

material space—the actual space and its forms and objects; *representational space*—the knowledge that is produced about space and its production; and *lived space*—the emotional experience of space and the subjective practices that are attached to the space. Space, then, is at once a physical environment that can be perceived; a semiotic abstraction that informs both the common and the scientific knowledge with which people negotiate and produce space; and a medium through which the body lives out its life in interaction with other bodies.[31] In parallel with Lefebvre's triad, David Harvey proposes the deconstruction of space into absolute, relative, and relational space. He creates a matrix in which it is possible to locate various examinations of spatial practices and discourses (see Table I.1).[32] Known analyses of space tend to emphasize one of its facets over the others. For example, the treatments of de Certeau or Deleuze[33] of spatial practices are centered on the lived-relational facets of space, that is, the subjective, inner experiencing of space (represented by * in Table I.1). The problem with such treatments is that they remain in the realm of the mind and neglect to connect the production of space with its material facet. In analyzing the space of community gardens and the space of the city at large it is sometimes necessary to emphasize one facet of space or another, but an effort should be made to bring together the three facets and create a complicated (and politicized) understanding of space. The analysis of community gardens takes on this effort and captures the process of social change as it is manifested in all three facets of space (represented by the line in Table I.1).

Table I.1   **The matrix of the triadic facets of space**

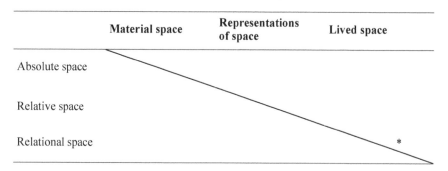

	Material space	Representations of space	Lived space
Absolute space			
Relative space			
Relational space			*

---

31   Lefebvre 2003.

32   Harvey, D. 2006b. *Spaces of Global Capitalism: Towards a Theory of Uneven Geographical Development*. New York: Verso. His deconstruction of the triadic facets of space is elaborated on in Chapter 5.

33   De Certeau, M. 1988. *The Practice of Everyday Life*. Berkeley: University of California Press on the practice of everyday life; Deleuze, G. 1994. *Difference and Repetition*. Trans. Paul Patton. New York: Columbia University Press.

From Property Rights to the Right to the City

Another consequence of modern transformations of space is that space is no longer a product of a participatory and an ongoing process—the human process of its production—but is being dropped on earth from the abstract levels of rational planning and design.[34] Once produced, space is being presented as a final product, not lending itself to innovations, modification, and participation on the part of its human users.

The implications of centralized, modernist planning and production of space are far-reaching. Not only is the space we live in not intended to accommodate the full scope of human needs, and in many cases stands in strong conflict with these needs, but also people are cut off from the opportunity to "inhabit" the urban space.[35] To "live" in the city is quite different from "inhabiting" the city. Lefebvre uses the metaphor of a seashell that produces its own shell over time to describe the organic relations between inhabitants and their habitat. If the creature is separated from its shell it is rendered nothing more than "something soft, slimy and shapeless."[36] Inhabiting indicates a dynamic process between people and their space that involves participation of the former in the production of the latter, participation that implies a sort of unity between the two. While "working in" or "being in" the city implies a passive, consumerist position of occupying space, "inhabiting" the city constitutes urbanites as active agents, taking part in the city's social life and production. Detached from the creation of their own shell, or the opportunity to inhabit, people lose their creative capacities and are left to operate in a world created by others. They are also detached from the idea that they have the power to produce their own space or that they may have a right to do so. Therefore, the "loss of inhabiting is a political, social, and aesthetic loss," deeming the habitat—the city and space—a mere container of life.[37]

The organic unity between people and their space, implicit in residents' dynamic and creative role in the process of the production of space, produces not only meaningful spaces but meaningful beings as well. The urban environment— where "life is lived in the rhythm of capital and by the logic of commodity"— is where this unity has suffered the most degradation.[38] Everyday urban life is colonized by the dominant mode of production that induces consumption as a unified life activity and alienates everyday life from its cultures and history.[39] But

---

34   Lefebvre 1991.

35   Lefebvre, H. 1996. "Part II: *Right to the City*," in E. Kofman and E. Lebas (eds), *Writings on Cities*. Oxford: Blackwell, pp. 63–264.

36   Lefebvre, H. 1995. *Introduction to Modernity*. New York: Verso, p. 116.

37   Merrifield 2005, p. 68.

38   Highmore, B. 2002. *Everyday Life and Cultural Theory: An Introduction*. New York: Routledge, p. 32.

39   Lefebvre, H. 2002. *Critique of Everyday Life: Foundations for a Sociology of the Everyday* (vol. 2). New York: Verso.

alienation can be transformed exactly where it is most intensified. It is possible to de-alienate everyday life by the overturning of cultural values, the constitution of another way of being, and by the reconstitution of the organic unity between people and the environment.[40]

Against the backdrop of the theory of space and the theory of the everyday— this work examines community gardens in New York City as a counter-force within the urban environment which strives to resist and transform the prevailing social and spatial environment. Since space not only reflects social conditions but also constitutes them, it is a site "through which identity politics, citizenship, and alternative political agendas are articulated and struggled over.[41] As the dominant production of space dramatically influences the life of urbanites, the struggle for change, for the appropriation and reclamation of space for differential usages, is articulated in terms of the organic unity between people and their environment, civil liberties, and the "right to the city."

This work follows the efforts of community gardeners to reassert and redefine their right to the city. These rights claims are construed through the "struggle to define and appropriate" space, that is, "the right to claim presence in the city, to wrest the use of the city from privileged new masters and democratize its spaces."[42] The chapters reveal how gardeners establish their right to participate in the production of the city through spatial practices over material space, discourse on space (that is, representational space), and the integration of space into everyday life (that is, lived space). Any progress with rights to the city is a step toward the *right to the oeuvre*—that is, the right to full participation in civilization; and toward the constitution of a radical new social subject—the *citadin*.[43] "Citadins" are at the core of Lefebvre's visionary new urban politics—urban politics of the inhabitants. They endow urban inhabitants with a new rightful status in the city, which will come about through:

> a radical restructuring of social, political, and economic relations, both in the
> city and beyond ... [that] reorients decision-making away from the state and
> toward the production of urban space ... [and] restructures the power relations
> that underlie the production of urban space ... [by] shifting control away from
> capital and the state and toward urban inhabitants.[44]

---

40   Ibid.

41   McCann, E. 2002. "Space, Citizenship, and the Right to the City: A Brief Overview." *GeoJournal* 58: 77–9, p. 77.

42   Isin, E. (ed.). 2000. *Democracy, Citizenship and the Global City: Governance and Change in the Global Era.* London: Routledge, p. 14.

43   Lefebvre 1996, pp. 63–264.

44   Purcell, M. 2002. "Excavating Lefebvre: The Right to the City and its Urban Politics of the Inhabitants." *GeoJournal* 58: 99–108, pp. 101–2.

Community gardens are sites where efforts are made to articulate new politics and demand new rights, rights to the city. Expressions of this new politics are found not only in the alternative usage of the space itself (for example, the dominancy of the use value of space) but also in the efforts of participants to protest urban policy and decision making and suggest an alternative vision for the city. The rights that are promoted by community gardeners by praxis and discourse advocate for a city that does not privilege the market and its rationale but for a city that belongs to its users—the inhabitants.

The Chapters

The first chapter, *A Garden in the City*, presents the history of community gardens in the US during the twentieth century and demarcates the contemporary phase of community gardens in the unique spatio-political context of New York City. It provides a brief overview as well as a quantitative summary of community gardens in New York City today.

The following chapters are organized in three parts paralleling the three analytical levels: individual, collective, institution. Part I: *Cultivating a New Individual: Life, Needs, Desires* focuses on the experiences of individuals in community gardens. It presents the more personal, experiential, and psychic aspects of everyday life that are being altered through the production of community gardens. These actions of individual place-making are conceptualized as the resilience practices of the traditionally marginalized urban residents in order to make a life in the city.

This first part portrays the building blocks that constitute the gardens as a differential space in relation to their surrounding urban environment. Based on these building blocks, different social relations and spatial practices are formed. With this delineation it is possible to bring to the surface the dialectical process of politicization. The individual "blocks" are a layer of experiences, practices, and consciousness from which extra-individual layers of social actions and consciousness are developed. This part reveals how the 'political' germinates in the most individualistic aspects of life.

The building blocks of individual experience of the garden are presented in a certain order. First are three facets of direct, everyday life experience that are constituted in and enabled by the interaction with the space of the garden: the constitution of the garden as a connective tissue between past landscape of gardeners and their present environment; the reintegration of festivity, spontaneity, and uncommodified celebration and creativity into life; and the reinforcement between inclusiveness of multiple identities and needs and the process of identification with space. Together these building blocks tell the story of how residents reconstruct their living environment in their own image (Chapter 2: *A Place in Their Image*).[45]

Next are some of the significant implications or consequences of the deepening relations between residents and their living environment (through direct, positive, and supportive interaction with it). One such implication is the activation of one's sense of ownership (psychological ownership) and experience of control over the environment which then reinforce the emotional significance of the environment. Two types of interaction between gardeners and their space are identified: proactive and reactive. These approaches to spaces characterize the initial level of agency of residents when they became gardeners and enable us to track the development of residents' agency through spatial and social practices (Chapter 3: *Determining Space, Creating Opportunities*).[46]

Part II: *The Spring of the Commons* focuses on the gardens as a social space, where actions and experiences are group-dependent: collective experience and production of space. It shows the processes through which space is socially produced and analyzes its products. Discussing these collective actions of community gardeners as part of the *reworking* practices of underprivileged urban residents to redress some of their local problems, this part begins to delineate the transition that happens in the garden away from survival practices and toward more explicitly political consciousness and practice.

The collective of community gardeners is formed both symbolically—through their shared narratives and history, and through their shared position within the social order and urban power structure—and actually through collaborations. As a diverse collective characterized by relatively loose ties, it is saturated with conflicts and internal challenges. Nevertheless, the process of collective production of space lays out the conditions needed for this collective to evolve and grow. The formation of a collective is significant for the development of both ideas and actions that perpetuate the production of space (Chapter 4: *The Formation of the Collective*).

Once a collaborating community of residents is formed, gardens become a rather unique space in the city—a space of actually existing commons. The gardens function as a collective resource of open space, food supply, and cultural and artistic space—and as a source of new skills and knowledge. Examining expressions of actually existing commons as they are manifested in and through the gardens not only illuminates the unique role of community gardens in constituting alternative urban experience, new identities, social relations, and political practices but also helps us envision the commons in the neoliberal, modern city. The commons are

---

45 A version of this chapter was published in Eizenberg, E. 2010. "Remembering Forgotten Landscapes: Community Gardens in New York City and the Reconstruction of Cultural Diversity." in T. Fenster and H. Yacobi (eds) *Remembering, Forgetting and City Builders*. Farnham: Ashgate, pp. 7–26.

46 A version of this chapter was published in Eizenberg, E. 2012. "The Changing Meaning of Community Space: Two models of NGO management of community gardens in New York City." *International Journal of Urban and Regional Research* 36(1): 106–20.

the product of social production of space that affords an alternative life experience in which a diversity of cultures, aesthetics, ideas, and actions is celebrated and where people can participate in the process of production as non-consumerist social actors (Chapter 5: *Actually Existing Commons*).[47]

In Part III: *Reaped Politicization*, community gardens are examined as a social movement that is involved in claims for the right to the city. Community gardening, a sporadic and very local phenomenon, was going through a process of consolidation and institutionalization toward becoming a proactive social movement. The institution of community gardens is an assemblage of organizations that are tied together, though somewhat loosely. Nevertheless the institution has some main pillars, several defined and agreed upon goals and strategies of struggle, and inherent tensions between the grassroots nature of community gardens and the process of their institutionalization. The institution of community gardens articulates the discourse and orchestrates the practices of community gardens in New York City. This discourse and these practices suggest that community gardeners developed an oppositional consciousness, and their actions are directed toward broader social problems. The institution of community gardens carries on a resistance and a struggle for a new politics of the inhabitants (Chapter 6: *Rooting Politics: The Institution of Community Gardens*).

Moreover, the collective of community gardeners in New York City through its vision, space, and institutions produces an arena for the development of political consciousness that facilitates the emergence of a different type of urban resident—what I am calling "organic residents." A significant interaction with one's living environment must include a strong personal attachment, sense of ownership, and high level of control over the environment; an organized collective action; and the creation of an institution that reproduces the collective and articulates its agenda (Chapter 7: *Setting the Ground for Organic Residents*).

Finally, concluding remarks suggest a broader relevance of community gardens through three points of reference: urban politics, urban nature, and urban planning and community development.

---

47 A version of this chapter was published in Eizenberg, E. 2012. "Actually Existing Commons: Three Facets of Space of Community Gardens in New York City." *Antipode* 44(3): 764–82.

Chapter 1

A Garden in the City: A History of Uneven Urban Development and Redevelopment

The History of Community Gardens in the United States

Urban gardening has a long history in the United States. By the nineteenth century new immigrants and workers were concentrated around growing industrial centers in the US in poor conditions of overcrowding, illness, crime, and stress. These newly formed "disaster areas" were viewed as a cultural failure by their contemporaries, who sought ways to help people cope with this new way of life. Solutions were inspired by seventeenth- and eighteenth-century romanticism which conceived of nature as having recuperating capacities. Advocates asserted the need for greenery and open space in order to produce a more wholesome, industrious citizenry by virtue of simple exposure to natural stimuli. Simple manipulation of the landscape can influence thoughts, experiences, and behaviors of urban inhabitants so they can recuperate from the hardship of urban life through access to nature in the city. Hence, in the mid- to late nineteenth century many cities established large-scale parks. In addition, late nineteenth-century reformers in cities such as Detroit, Philadelphia, and New York initiated vacant-lot cultivation associations as poor-relief programs, while educators initiated the school gardens program. The vacant-lot cultivation associations, or community gardens, were viewed as embodying both material and spiritual values in an era of adjustment to life in industrial cities. They aimed to provide relief for the poor in a relatively dignified manner, as well as to serve as a melting pot where immigrants and natives could get together and facilitate the learning of the "American way."[1]

In the years to come community gardens gained and lost popular interest in parallel with economic crises. In the two World Wars of the twentieth century, urban lot cultivation was publicized as a symbol of patriotism; urban residents participating in the gardens programs were perceived as "plant[ing] for freedom"; and the gardens were regarded as "war gardens" in World War I and as "victory gardens" in World War II. Particularly in the World War II, when provisions from rural areas were scarce, city people were called upon to cultivate everywhere and

---

1 Bassett, T. 1979. "Vacant Lot Cultivation: Community Gardening in America, 1893–1978." Unpublished manuscript, Department of Geography, University of California, Berkeley; Lawson, L. 2005. *City Bountiful: A Century of Community Gardening in America.* Berkeley: University of California Press.

to provide fruits and vegetables for themselves in order to win the war. After World War II, however, community gardens gradually disappeared from the cityscape until their rejuvenation in the mid-1970s.[2]

For over a half a century, then, community gardens were "top-down" produced spaces aimed at socializing, integrating, assimilating, and molding those who were considered to be maladjusted (or had adjustment difficulties)—the poor, immigrants, and children—to the new American project of industrial capitalism, or as part of a united home-front in times of war. However from the 1970s to the present, community gardens have represented a dramatically different "bottom-up" phase in the history of urban community gardening. Historically, this book focuses on the contemporary phase of community gardens, and geographically on community gardens in New York City. Some characteristics of the new conceptualization of community gardens in their contemporary phase can also be identified in other major urban settings in the United States (and elsewhere) that are rich with community gardens—such as San Francisco, Seattle, Philadelphia, and Boston, but other characteristics are unique to community gardens in New York City.

Without differentiating between locations, there are three main reasons for identifying and conceptualizing this new phase of community gardens. First, the backdrop for contemporary community gardens is significantly different as cities underwent an extensive restructuring process guided by a new rationale and ideology. Second, this phase of community gardens is characterized by a bottom-up, grassroots production of the gardens, led by urban residents rather than by reformers, educators, or war campaigners. And third, the new phase of community gardens is characterized by continual assaults from powerful urban forces, assaults which the gardens not only endured successfully (at least partially) but also impacted and changed the contemporary urban discourses and practices.

The Special Context of New York City

Community gardens in their current phase can only be understood within the specific socio-political context of their location, and New York City is an especially unique context for the gardens. Among major US cities, New York has been one of the most fertile breeding grounds for neoliberal restructuring since the beginning of the 1970s.[3] Three significant developments can be identified since the 1970s that support this claim. First, the upper class and its wealth returned to the inner city— turning back the tide of the white flight to the suburbs in the 1950s and 1960s— and has come to dominate the space. It was suggested, for example, that the fact that a millionaire—Michael Bloomberg—was elected as the city's mayor was not

---

    2   Lawson 2005.
    3   Harvey, D. 2006b. *Space of Global Capitalism: Towards a Theory of Uneven Geographical Development*. New York: Verso.

coincidental but symptomatic of the rising power of the upper class in New York.[4] The second development came about as a result of the fiscal crisis of the 1970s. The crisis facilitated parallel and related trends of diminishing the power of labor unions and of quick urban deindustrialization, enabling the city to become the center for the financial sector, a global city. The third major development was the shift in city planning schema which guided the redesign of growing portions of urban space as tourist destinations. Marketing Times Square as the symbol of New York City epitomizes this change. Another manifestation of this new planning schema is the increased coverage of the city's squares with grass and flowerbeds: a new configuration of public space which does not accommodate political gathering and protest and is more controllable and sterile.[5]

While such trends of urban restructuring are not unique to New York City, they were more dramatic there because of several characteristics. New York City stands out in terms of low open space per capita among high-density cities, and although its expenditure on open space and parks per resident grew in the last decade, it still lags behind other big cities in the US.[6] In addition, after 2006 the city was left with no undesignated parcels of land in its inventory, making it very difficult for the city to create new open spaces.[7] The city, with its high rents and the privatization and suburbanization of public space, no longer caters to the poor and the marginalized.[8] Its malignant design of urban space directly attacks the poor and the working class and results in their almost complete displacement from main urban centers. Current planning and design of space not only attacks these groups of residents as a social class but also the spatial expressions of their cultures. Community gardens exemplify a kind of a counter-reaction that such a geopolitical context generates.

The economic crisis of the early 1970s marks a watershed in the history of the city, unleashing dramatic urban transformations. The coupling of a period of public and private disinvestment with the expansion of immigrants' neighborhoods resulted in landlord-abandoned property, collapsed or intentionally dismantled buildings, and general decay in many of New York City's neighborhoods. Rubble-strewn vacant

---

4   Harvey, D. 2006a. "Public Space and the Uses of the City." Paper presented at the conference on Refashioning Urban Spaces in Paris and New York for the 21st Century, New York University, April 29.

5   Mitchell, D. 1995. "The End of Public Space? People's Park, Definitions of Public and Democracy." *Annals of the Association of American Geographers* 85: 108–33. Mitchell explains how public spaces are beng redesigned in order to accommodate very specific activities and forms of usage that suit certain desirable groups while driving out other groups.

6   http://cloud.tpl.org/pubs/ccpe-city-park-facts-2011.pdf (updated for 2009; accessed December 2011).

7   Department of Housing Preservation, and Development (HPD) representative reports at the 2006 Annual Gardeners' Forum, April 22, 2006.

8   Harvey 2006a.

lots multiplied in the most devastated parts of the city, and turned into hotbeds for local problems of sanitation, violence, crime, and lack of recreational amenities.[9]

Community gardens evolved at this historical point out of the efforts of residents to initiate community development, most significantly in neighborhoods that suffered the most from blight and decay (such as the Lower East Side and Harlem in Manhattan; the South Bronx; and Bedford-Stuyvesant and East New York in Brooklyn). What started out as residents planting flowers on the edge of lots and in tree pits, and throwing "seeds bombs" over fences, became the efforts of transforming rubble-strewn vacant land into communal green spaces for recreation, gardening, and even small-scale urban farming.

The Liz Christy Community Garden in the Lower East Side is allegedly the first community garden, founded in 1973 on the corner of Houston Street and Bowery. A group of urban activists, later organized as the Green Guerillas,[10] inspired and helped other groups to follow suit. And they did, many of them in the poorest neighborhoods of the city. Residents' groups leased urban lots from the city for $1 per year and made them into community gardens. The city enjoyed these grassroots efforts as a cost-free community revitalization program.[11] Operation Green Thumb, a city agency, was inaugurated in 1978 to oversee and encourage community groups to develop community-managed open-space projects. It became an agency of the Parks and Recreation Department in 1995.[12] At the same time, gardeners' coalitions were

---

9   Francis, M., L. Cashdan, and L. Paxson. 1984. *Community Open Space: Greening Neighborhoods through Community Action and Land Conservation.* Washington, DC: Island Press.

10   Green Guerillas is a nonprofit organization helping community gardens in the city. It focuses on facilitating gardens' organization and works closely with neighborhood coalitions (most notably, but not exclusively, East New York Gardens Association, Bedford-Stuyvesant Gardens Coalition, and Harlem United Gardens). Green Guerillas helps groups develop operational procedures, articulate their agendas and goals, and pursue them. The organization runs several initiatives to engage youth in community gardens, such as the Youth Mural Project. In addition, Green Guerillas helps gardens in need through materials and plants provision and construction assistance or guidance. Towards the end of the first decade of the twenty-first century, Green Guerillas was involved with Ocean Hill/Bedford-Stuyvesant residents in an effort to clean large vacant lots in their neighborhoods and transform them into an urban farm, a community garden, and an education center (interview with Green Guerillas representatives on September 29, 2004 and November 8, 2005).

11   Schmelzkopf, K. 1995. "Urban Community Gardens as a Contested Space." *Geographical Review* 85(3): 364–81.

12   According to its Executive Director, Green Thumb is the largest urban gardening program in the nation. The agency serves over 600 community gardens and 20,000 garden members by providing the required licenses (in which gardeners commit to general rules of opening hours and membership), training in a plethora of garden-related subjects, and the provision of material and technical support. Green Thumb is funded by federal grants and from 2007 also by tax levies. Its operating budget for 2006 was $235,000 (interview November 15, 2006).

established in New York, as well as the national American Community Gardens Association (ACGA) that was established in 1983.[13]

Despite this blooming and embracing of urban gardening, since the 1980s, as the financial and real estate market began to recover and reinvestment became more feasible, city authorities changed their attitude towards community gardens. The economic crisis was followed by two consecutive waves of gentrification. The first wave was in the late 1970s and 1980s and the second in the early 1990s. Both waves were characterized by increasing levels of state and municipal intervention bringing high-end businesses and populations into the inner city while inevitably facilitating the displacement of existing residents.[14]

As a result, from the early 1980s community gardeners were forced into ongoing confrontations with the municipality and real estate entrepreneurs who partnered them in a quest to hand over "vacant" land for profit-driven redevelopment. This confrontation reached its peak when Mayor Rudolph Giuliani listed 126 gardens for an auction scheduled for May 1999, planning to sell off hundreds more and publicly denouncing the gardens as an obstacle to a free market economy.[15]

In over a century of community gardening across the US, gardens had periods of broad popular support and periods when they were forgotten in minds and praxis, but only on rare occasions they were considered as a permanent land use designation.[16] However, the assault by the Giuliani Administration, aiming to bring a vibrant period of community gardens to an end, had the opposite effect: it changed the temporal status of community gardens in New York City. Threats to the gardens generated a massive public outcry, bringing together gardeners, activists, neighbors, and local green organizations in a struggle against efforts to eradicate the gardens. People marched in the streets, protested in front of City Hall, filed lawsuits against the city, and chained themselves to garden gates to stop approaching bulldozers. In February 2000 the then New York State Attorney General, Eliot Spitzer, signed a court restraining order requiring the city to stop all destruction of gardens. Though it was already too late for some community gardens,[17] the order prevented the destruction of many other gardens. Meanwhile, two nonprofit organizations—the Trust for Public Land (TPL) and the New York

---

13   Francis, Cashdan, and Paxson 1984.

14   Hackworth, J. and N. Smith. 2001. "The Changing State of Gentrification." *Journal of Economic and Social Geography* 92(4): 464–77.

15   Schmelzkopf, K. 2002. "Incommensurability, Land Use, and the Right to Space; Community Gardens in New York City." *Urban Geography* 23(4): 323–43. In a confrontation with gardeners protesting against plans to auction off many gardens, Mayor Giuliani addressed the activists, dressed as flowers and bees: "This is a free-market economy. Welcome to the era after communism." Shepard, B. and R. Hayduk. 2002. *From ACT UP to the WTO: Urban Protest and Community Building in the Era of Globalization.* New York: Verso, p. 200.

16   Lawson 2005.

17   Several gardens, such as the famous and vital Esperanza community garden in the Lower East Side, Manhattan, were already evacuated.

Restoration Project (NYRP)—negotiated with the city and purchased about 120 gardens that had been on the auction list (for $4.2 million).

In September 2002 a settlement with the city, now headed by Mayor Bloomberg, was signed. The Gardens Settlement decreed that the gardens could no longer be considered "vacant lots" in the process of auctioning land for development. Now, a "Garden Review Statement" would document the garden's history and be presented to community boards, city council members, and the City Planning Commission as part of the Unified Land Use Review Procedure (ULURP) before any approval of development. In addition, 193 community gardens were granted preservation status under the auspices of Green Thumb. These included 100 school gardens that were under the jurisdiction of the Department of Education and 86 gardens that were already part of the Parks and Recreation Department before the settlement. Another 198 gardens were transferred as part of the settlement from various jurisdictions, including the Department of Housing Preservation and Development (HPD), to the Department of Parks and Recreation and were also preserved. The 152 community gardens that were appropriated by the HPD and were slated for immediate or future development were promised "a new 'garden review process' that requires the City to offer alternative properties (if available) to the affected gardeners, and to provide information about the garden to the gardeners and elected officials before land use decisions are made."[18]

The history of community gardens in New York City is one of a struggle waged by underprivileged residents of the city to improve their neighborhoods and everyday life conditions. Whether struggling to establish the gardens in the midst of urban decay or later on struggling to preserve them in the face of powerful market forces and unsupportive municipality, the gardens can be understood as an alternative force and counter-reaction to the prevailing urban order.

In the devastated urban context of the 1970s, when the first gardens appeared in New York City, community gardens were marginal spaces on the spatial maps of capitalism. They can be understood as heterotopias in the sense that they negated, and to some extent reversed, the destruction that prevailed around them. This general urban destruction of the 1970s can be seen as part of a process of what Schumpeter termed *creative destruction*: planned (or not) destruction of cities (or parts of cities) in order to rebuild them anew as part of a capital accumulation strategy.[19] The inauguration of community gardens, right in the midst of a harsh disinvestment period, hindered those destructive forces. In the same vein, if this destruction was supposed to set the ground for the creative implementation of neoliberal urban principles, then the gardens also hindered the *creative* phase of this process (usually referred to as "development" or "progress," indicating

---

18   NYC Community Gardens Agreement, September 2002, http://www.ag.ny.gov/ bureaus/environmental/pdfs/community_gardens_agreement.pdf (accessed November 2011).

19   Harvey, D. 2003. "The City as a Body Politic." In J. Schneider and I. Susser (eds), *Wounded Cities: Destruction and Reconstruction in a Globalized World*. New York: Berg, pp. 25–44.

a universal benevolence) by protecting pockets of urban land from privatization and promoting a strong public sphere in the face of increasing individuation. They were interjecting an alternative rationale to the profit-driven market rationale that dominates the process of urban decision making.

The 2002 Gardens Settlement gave gardeners eight years, until its expiration in 2010, to figure out their new status and role in the city and to act accordingly. Apart of some glimpses to the past, this book is nested in the first decade of the second millennium when resident-gardeners were routinizing their activism. New rules for gardens in New York City, replacing the settlement, were announced in the fall of 2010 after long deliberations between the city, community gardeners, green organizations, and state representatives. No immediate changes were issued for the status of community gardens in the city, both a positive and a negative declaration. Preserved gardens would continue to be preserved as long as they comply with Green Thumb licensing and the garden review process in order to maintain good communication with community boards regarding the gardens. In addition, the new agreement stated that, although not required to, the city may create new gardens and those would be subjected to the same rules of preservation. Moreover, gardens were not and would not be mapped as parkland; they are not perceived as an integral part of the landscape and may be snatched away if the community fails to license them.

Uneven Urban Development

As rationalized and commodified as urban land may be, it was difficult to come up with an accurate number of community gardens. There are several reasons for this difficulty. The grassroots nature of the gardens disposes them to appear spontaneously on the urban scene and in some cases to operate without licensing from city authorities. For a few years after their first appearance, between 1973 and 1978, community gardens were a "no-man's-land"—undocumented and unsupervised appropriations of land. After 1978, with the inauguration of Operation Green Thumb, there was a gradually developing effort to legalize, supervise, and more recently standardize community gardens. Nevertheless, some gardens still evade private or municipal interest or control. The spontaneous and dynamic nature of the gardens makes them more difficult to track down. Even after the 2002 New York City Community Gardens Settlement that ordered these grassroots spaces into more conventional land designation, the pool of community is still in flux; gardens are being developed, relocated to new sites, or integrated into other existing gardens. In addition, efforts to maintain a tracking record of community gardens are very recent and are therefore not comprehensive and sophisticated enough. Green Thumb, the body that oversees all the gardens in the city, only began to compile a systematic database in 2005. The information from this database (updated to October 2006) is in some disaccord with another database edited by the Council on the Environment of

New York City—CENYC[20] (updated for February 2008). Nevertheless, these two resources are the only citywide resources available and I therefore use them in this book, but I use them with caution.

Why was the city not developing a coherent database after so many years of overseeing the gardens? Mere negligence or lack of resources, both may be the answer; but there might also be some ideological explanation. Until the settlement of 2002 gardens were perceived as "vacant lot" by the municipality and there was no incentive to list them separately. Ambiguity of information regarding the gardens helped avoid friction in processes of urban planning and development. After the settlement there was an obvious need for an orderly listing of the preserved gardens but that was developing slowly. According to Green Thumb's representative,[21] there were efforts to present the 2002 Gardens' Settlement as an impressive public victory. The administration wanted to present the new Mayor, Michael Bloomberg, as responsive to the public rather than to private business needs. To do so the number of preserved gardens was exaggerated in press releases and mayoral presentations, and then crunched to fit the reality. Gardens moved from one listing to another and a comprehensive dataset was not compiled.[22]

Since the 1970s then, urban community gardens have been initiated, produced, and maintained by groups of residents rather than as a top-down program of reformers or educators. Members of contemporary community gardens decide on the agenda, purpose, and usage of the space according to their worldview, culture, and needs. In many ways, contemporary community gardens represent different ideas of urban land use in which the "space ha[s] rarely been planned as part of development but happen[s] after the fact, often on deserted, derelict or otherwise unused land."[23]

The vast majority of community gardens in New York City are concentrated in Brooklyn (332 gardens), Manhattan (200), and the Bronx (158).[24] However,

---

20   The Council on the Environment of New York City (CENYC) was established in 1970 as a nonprofit organization that is also part of the mayor's office. It is therefore a private-public partnership. It runs three key programs: urban green markets; environmental education programs (primarily for high schools); and the Open Space Greening Program, which is responsible for community gardens, among other things. Under the Plant-a-Lot initiative, CENYC helped gardeners create over 50 community gardens across the city and continues to help create one to two new gardens each year. Gardens in the Plant-a-Lot initiative receive ongoing support with maintenance, organizing, and events. CENYC also operates the "grow track," a program which helps provide tools and supplies to the gardens, and that conducts a yearly plants sale providing plants for over 100 garden groups. Another major project under the auspices of CENYC is the web-accessible mapping of community gardens. Finally, CENYC delivers educational workshops for gardeners (interview December 5, 2005 with a representative of the Open Space Greening Program at CENYC).

21   Interview, November 15, 2006.

22   One way to "fix" the numbers was by moving school gardens from the Department of Education to the list of community gardens under the Parks and Recreation Department.

23   Lawson 2005, p. 2.

24   Based on CENYC's database, updated in February 2008.

this count includes gardens that are under the jurisdiction of the HPD and are designated for future development. The average size of a garden in the city is about 5,000 sq. ft (460 sq. m). According to incomplete but still indicative information on the land coverage of gardens, 184 Manhattan gardens (of the 200 total) cover 17.4 acres in total; 312 community gardens in Brooklyn (of the 332) cover a total of 40.3 acres; 149 community gardens in the Bronx (of the 158) cover 23.4 acres; and 39 gardens in Queens (out of 46) cover 9 acres. Thus the aggregated acreage of community green space in New York City provided by community gardens is at least 90.1 acres.[25]

There is a specific pattern to the distribution of community gardens in New York City: several urban areas with some shared characteristics encompass high densities of community gardens. Examining socio-economic characteristics such as race, median household income, renter-occupied units, and median gross rent in areas of high concentration of community gardens—such as East New York, Bedford-Stuyvesant, and Ocean Hill in Brooklyn, the Lower East Side and Harlem in Manhattan, and Crotona and the South Bronx in the Bronx—draws a very clear and unequivocal picture regarding the population that uses the gardens and tends to them.

Community gardens are concentrated in areas with very high rates of African American and Hispanic populations, suggesting that gardeners citywide are predominantly African American and/or Hispanic (see Maps 1.1 and 1.2). For example, 87 percent of the population in East New York is either black or Hispanic (49 percent black and 38 percent Hispanic); 95 percent of the population in Bedford-Stuyvesant is black and Hispanic (77 percent black and 18 percent Hispanic); and Crotona neighborhood in the Bronx is 97 percent black and Hispanic (44 percent black and 53 percent Hispanic).[26]

The distribution of community gardens in the city does not only follow racial lines but also draws the complete picture of uneven urban development; the areas of the city that are saturated with community gardens encompass fairly low percentages of open space. Despite the high concentration of community gardens these districts fall short of the average ratio of open space per capita in New York (which was marked at 7.17 acres per 1,000 residents in 2000).[27] For example, in the Bedford-Stuyvesant neighborhood in Brooklyn with 65 community gardens, the ratio of open space per 1,000 residents was 0.27 acres in the same year. In South Bronx, with 40 community gardens, the ratio was 0.92 acres in 2000 and 2.08 acres in East Harlem with 53 community gardens.[28]

The distribution of community gardens juxtaposed with the distribution of median household income in three boroughs of the city indicates that the gardens are the creation , and serve, predominantly low-income residents. The gardens are

---

25   Ibid.

26   Based on 2000 census information.

27   Harnik, P. 2000. *Inside City Parks*. Washington, DC: Urban Land Institute.

28   This information was retrieved from http://www.oasisnyc.net, an open space mapping project of the Council on the Environment for New York City (updated May 2007).

1 - 150
151 - 500
501 - 1000
1001 - 2000
2001 - 3000
3001 - 6000
above 6000

Miles
0   1   2   4

**Map 1.1   The distribution of community gardens by black population in
                 three city boroughs**

Legend:

1 - 150
151 - 500
501 - 1000
1001 - 2000
2001 - 3000
3001 - 6000
above 6000

0   1.5   3   6   Miles

Map 1.2   The distribution of community gardens by Hispanic population in three city boroughs

☐	below 16000
▨	16001 - 25000
▨	25001 - 50000
▨	50001 - 75000
▨	above 75000

Map 1.3   The distribution of community gardens by median household annual income [US$] in three city boroughs

located in areas where the annual median household income is likely to be below $25,000. While the median household income in New York City stood at $38,518 for 2000, this parameter is even lower for areas with the highest concentration of community gardens (see Map 1.3). For example, in East Harlem (53 community gardens) and Central Harlem (40 gardens) the median household income for 2000 was $21,297 and $19,924 respectively; in the South Bronx (40 gardens) $16,000; and in Crotona, Bronx (41 gardens) $16,603.[29]

The correlation of gardens with economically distressed low-income neighborhoods can also be seen by juxtaposing the distribution of community gardens with median gross rent. Gardens are more likely to be located in areas with lower rates of median gross rent (see Map 1.4). While the median gross rent in New York for 2000 was $712, in zones of high concentration of gardens such as East and Central Harlem the median gross rent for 2000 stood at $463 and $483 respectively. In the South Bronx and Crotona the median gross rent stood at $349 and $429 respectively; and at $451 for Ocean Hill, Brooklyn.[30]

Finally, the distribution of community gardens in three New York City boroughs juxtaposed with the rate of renter-occupied units shows unequivocally that community gardens are located in areas that are overwhelmingly characterized by a population that rents, rather than owns, its housing. While the city's percentage of renter-occupied units is 69.8, in areas with a high concentration of gardens a much higher percentage of units are occupied by tenants (see Map 1.5). For example, over 90 percent of housing units in the Bronx community districts 1, 3, 4, 5, and 6 (all of which have a large number of gardens) are renter-occupied. A similar percentage of renters can be found in East and Central Harlem. Taken as a whole, the information reveals that community gardeners are predominantly low-income, African American and Hispanic renters living in highly dense areas of the city characterized by a high scarcity of open space and lower than average rents.

The uneven urban development that produced places of poverty and distress explains also the uneven distribution of community gardens in the city. Intensified disinvestments in the 1970s deepened residents' despair and they were looking for ways to help themselves and change their living conditions. Uneven urban development played out again later on in the decisions made between 1999 and 2002 regarding which gardens should be preserved and which should be slated for capitalist development. In some community boards gardens were lost in high numbers while others were able to preserve more of their gardens. East and Central Harlem, for example, lost over a quarter of their gardens compared with the Lower East Side, which lost eight percent of its gardens. The Ocean Hill neighborhood (CB16) in Brooklyn is an extraordinary example, losing 70 percent of the community gardens it used to have.[31] A combination of factors may

---

29   Based on 2000 census information.

30   Ibid.

31   Based on Green Thumb's database, updated in October 2006.

	below 500
	501 - 800
	801 - 1000
	1001 - 1500
	above 1500

Miles
0   1.5   3   6

Map 1.4   The distribution of community gardens by median gross rent [US$] in three city boroughs

Map 1.5   The distribution of community gardens by renter-occupied
units in three city boroughs

explain this unevenness. One is related to the differential capacity of gardeners to protect their gardens in their community boards, reifying traditional advantages of stronger population (with more money, education, and connection to politicians and professionals) to take better care of their interests. Another important factor is land speculation that directed the pressure towards certain areas.

A Common New York City Garden

Community gardening is a voluntary activity, but membership in a garden entails some obligations. These may include participation in a number of membership meetings, participation in workdays or an orientation day, and membership fees. Membership in the majority of the gardens does not require payment of membership fees (in 438 out of a total of 650 gardens). In other gardens, membership depends upon payment of dues ranging from $10 to $30 per year for a household. Members receive a key to the garden gate and the tool shed, and in some gardens a plot for individual cultivation. Most community gardens in the city are fenced-off lots and have gates that can be locked whenever garden members are not on site. The main reason for locking the gardens is insurance liability. The city sees garden members as sort of park rangers who are responsible for the safety of visitors and activities.

The majority of gardens (64 percent) have more than 10 household members, while children from the neighborhood use the gardens in the afternoon, on weekends, and in the context of their school nature education. Green Thumb requires the gardens to be open to the public for 10 hours per week and to have at least 10 listed members. Ninety four of the gardens reported having a waiting list of residents wishing to become members or plot owners.[32]

Most of the gardens contain vegetable plots—individual, communal, or both—but a few gardens have only ornamental planting. All gardens have some structures—such as tool sheds, chairs and benches, gazebos, pergolas, stages, small amphitheaters, and casitas—as well as amenities such as ponds, bird houses, solar panels, rainwater harvesting systems, and the like. In 153 gardens across the city there is an operating compost system.[33] A permanent art display, in the form of sculptures and murals, is another common characteristic of the gardens.

Even after the gardens' settlement of 2002 that for the first time ordered the pool of community gardens in New York City there is still a large diversity in the legal status of the gardens. The city's Department of Parks and Recreation is the largest organization for overseeing gardens in the city, in charge of 28 percent of the gardens; 15 percent of the gardens are under the two nonprofit organizations that purchased them from the city; a little less that 20 percent of gardens are under the HPD and the same percentage is under the Department of Education.

---

32   Based on CENYC's database, updated in February 2008.
33   Ibid.

There is still a significant number of gardens (10 percent) located on private lots, a handful of which are actually privately owned by the gardeners (Clinton Community Garden in Manhattan is one such example). Additionally a small percentage of gardens are under the jurisdiction of other municipal departments or New York State.

PART I
Cultivating a New Individual: Life, Needs, Desires

Conceptual Considerations of Individuals and their Environment

What makes people interested in their surroundings? What makes them break the individualistic pact and invest time and effort in non-private matters? What makes them believe that they can have an impact on the public sphere? And what makes them try to induce change? As the regime of the individual with its values of self-fulfillment and personal achievements and gains becomes more prevalent, these questions get harder to answer.

Environmental social sciences use various concepts in order to understand the significant and complex interactions between urban residents and their living environment, interactions that evolve into diversified cultural landscapes. Concepts like place attachment,[1] place making,[2] sense of place,[3] identification with the environment, and place identity[4] enabled and enriched the discussion on human–environment relations—though nuances between them are not fully explicated. With little differentiation between them, they are being used interchangeably and remain somewhat vague in their explanations of the development of these sensations and psychological structures.

Place attachment "involves an interplay of affects and emotions, knowledge and beliefs, and behaviors and action in reference to a place," whereas *place* refers to "space that has been given meaning through personal, group, or cultural processes."[5] This definition does not indicate directionality: is place attachment the result of these processes, the processes themselves, or both? How are all these factors (emotions, beliefs, knowledge, etc.) developed in relation to or attached

---

1 Altman, I. and S. Low (eds). 2002. *Place Attachment*. New York: Plenum Press.

2 Lawrence, D. 1992. "Transcendence of Place: The Role of La Placeta in Valencia's Las Fallas," In I. Altman and S. Low (eds), *Place Attachment*. New York: Plenum Press, pp. 211–30.

3 Hey, R. 1998. "Sense of Place in Developmental Context." *Journal of Environmental Psychology* 18: 5–29.

4 Proshansky, H., A. Fabian, and R. Kaminoff. 1983. "Place-Identity: Physical World Socialization of the Self." *Journal of Environmental Psychology* 3: 57–83.

5 Low, S. and I. Altman. 1992. "Place Attachment: A Conceptual Inquiry." In I. Altman and S. Low (eds), *Place Attachment*. New York: Plenum Press, pp. 1–12. p. 5.

to the environment? Moreover, it is hard to identify human agency as part of this interplay. Rather, the impression is that the human subject has a relatively passive role in the development of place attachment. Although place attachment implies a process, it is often treated as static, as something you have or do not have.[6] A typology of place attachment refines the essences of human–environment bonding: genealogy, loss and destruction, economic, cosmological, pilgrimage, and narrative.[7] While this typology complicates the definition of place attachment by adding individuals' personal history to the process, the human subject nevertheless seems as succumbing to historical structures and is hence stripped of agency.

Sense of place relates somewhat different qualities to the human–environment relations and makes more explicit the social and geographical context of place bonds and the sensations that are evoked by places (for example, aesthetics, control, and feeling of dwelling).[8]

In contrast with the lack of agency implicit from place attachment, the ecological psychology exponents William James and James Gibson emphasize a pragmatic approach of individuals towards the environment. Their approach explains the interaction of individuals with the environment as a series of pragmatic decisions that propel and define their relation to it. Pragmatism assumes the unchallenged capacity (maximum agency) of individuals to make their best decisions for proceeding onwards. There are ample opportunities—affordances— that individuals can extract through their interaction with the environment.[9] However, such an understanding of the interaction with the environment completely overlooks socially constructed impediments such as power relations, conflicts, and physical constraints.

The concept of *place making* emphasizes a process of mutual reinforcement between people and the environment through which place is endowed with special meaning. The investment of affect, cognition, and practices in the place as well as the feedback from the environment to these investments constitute the development of meaning, attachment, sense of place, and identification with the environment. This definition adds "generative principles of action rather than being attached to place as an object."[10] This way of looking at the interaction with the environment adds a new layer; it is not only that the interaction of individuals with

6   See for example Brown, B., D. Perkins, and G. Brown. 2003. "Place Attachment in a Revitalizing Neighborhood: Individual and Block Levels of Analysis." *Journal of Environmental Psychology* 23(3): 259–71; Vaske, J. and K. Kobrin. 2001. "Place Attachment and Environmentally Responsible Behavior." *Journal of Environmental Education* 32(4): 16–21.

7   Low, S. 1992. "Symbolic Ties that Bind: Place Attachment in the Plaza." In I. Altman and S. Low (eds), *Place Attachment*. New York: Plenum Press, pp. 165–84.

8   Hey 1998.

9   Heft, H. 2001. *Ecological Psychology in Context: James Gibson, Roger Baker, and the Legacy of William James's Radica Empiricism*. Philadelphia: Lawrence Erlbaum Associates.

10   Lawrence 1992, p. 215.

the environment reveals its affordances[11] but also new meanings and opportunities are *produced* through the interaction and change both the environment and the person who acts upon it.[12]

While the concept of place making highlights the dialectic character of our relations with the environment through assigning a more active role to human agents, it still does not clarify motivations, goals, decision making, and the manner in which social conflicts intervene and shape the interaction.

The two concepts—"place attachment" and "place making"—describe facets of the meaningful linkage between people and the environment but downplay the political potential of spatial processes. The dichotomy between the outside physical world and historical reality on the one hand and inner processes of cognitive decision making on the other hand persists in psychologically centered explanations for human–environment relations. This disintegration between structure and agency in contemporary conceptualization makes it difficult to connect with the discourse on urban politics, transformation, and resistance. In a similar way, the latter discourse relies on the theorizing of social movements and collective actions, and for the most part ignores the contribution of individuals to these processes.

Integrating agency *and* power into the discussion on human–environment relations opens up a path in which significant and meaningful interactions not only register as significant places for individuals[13] but also render urban residents as social actors. For the poor and marginalized residents this social role entails their engagement with critical spatial discourse and practices of struggle through appropriation of space from the dominant forces.

In an effort to overcome the discrepancies and caveats of currently used concepts, a different concept should be all-encompassing enough to address the missing components of the other concepts and yet retain some openness for development and modification. The concept of *social production of space* seems to be more substantial and better suited to explain the intricacies of the interaction between humans and their environment, and specifically illuminating in the case of community gardens. Coined by Henri Lefebvre (1991), this concept understands space as being in an ongoing process of becoming. Social actors are constantly active in appropriating, using, and transforming space—which is thereby thoroughly social. Participants in

---

11 Gibson, J. 1979. *The Ecological Approach to Visual Perception.* Hillsdale: Lawrence Erlbaum Associates. Gibson suggests that the multiple properties and possible usage of objects (and space) are revealed through the interaction with them.

12 Lawrence 1992.

13 Manzo, L. 2005. "For Better or Worse: Exploring Multiple Dimensions of Place Meaning." *Journal of Environmental Psychology* 25(1): 67–86. Manzo suggests that significant places are not only places of positive affects but actually places of significant experiences which could be also negatively connoted or ambivalently connoted in consciousness. Therefore, identification with the environment is related to our complex experience of it.

the process have a conscious role regarding their space; interests, needs, conflicts (or in a word: power) are the nuts and bolts of this process.[14]

While space according to Lefebvre is predominantly a mechanism for social reproduction, it also encapsulates a promise: the promise of liberation from domination. Active appropriation and transformation of everyday environments by their inhabitants facilitates the reintegration of their lost history and their culture as well as art and aesthetics back into space. This production of space gives rise to diverse landscapes—differential spaces.

With this concept, Lefebvre introduces power, conflict, and interests to the understanding of the interaction of people with the environment. Although for Lefebvre the social production of space defines a social practice—a collective rather than an individualistic one—the concept is relatively open and undetermined enough to tolerate some integration of the psychological with the social, personal agency with structure, and needs and identity with power. Lefebvre's writings on the critique of everyday life provide clues for bridging these dichotomies. Through positing mundane experience and practice in the context of the socio-political realm he discusses how transformations in one's everyday life are keys to transforming the social and political realms.[15]

The two chapters of Part I present the mechanisms by which space becomes central to people's understanding of themselves and their everyday life. As such, space is constituted as an arena for practices and consciousness that are social and political, and hence go beyond the personal significance of the experience of space.

---

14   Lefebvre, H. 1991. *The Production of Space*. Oxford: Blackwell.
15   Lefebvre, H. 2002. *Critique of Everyday Life: Foundations for a Sociology of the Everyday* (vol. 2). New York: Verso.

Chapter 2
A Place in their Image

This chapter is the first attempt at drawing the theoretical continuum of the political development of residents that evolves out of the interaction between urban residents and community gardens. While focusing on the interaction of individuals with the environment, the chapter emphasizes the more personal aspects of the interaction that are less dependent on other people. In doing so the chapter illuminates those psychological processes of personal development that support and mobilize collective actions and political practices.

Three aspects of the interaction between gardeners and their gardens are discussed. First, the garden is presented as a place where individuals better recognize themselves in the physical environment through reconstructing elements of their past landscape. Past landscapes and spatial practices, elements of the environmental autobiography of gardeners, are being reconstituted by gardeners in the space of the gardens and in turn this reinforces a positive connection of gardeners with their living environment. Second, aesthetic and celebratory experiences that are afforded by the gardens proffer a unique contribution to the urban experience of participants. Third, the gardens are presented as constituting a significant component in the identity of the gardeners, one which is space- and community oriented. Finally, the chapter contextualizes issues of identity and resistance as they pertain to the individual gardener within the macro-elements of culture and citizenship in the modern city.

The Reconstruction of Past Landscapes

> People do not just exist in their own little worlds (personal geographies); however, they experience a shared picture of the world. Indeed, a sense of a solid shared world and a stable sense of self within that world are seen as essential for our psychic and physical survival.[1]

Memories of past landscapes and past practices are constitutive of the production of space of community gardens. Memories of natural landscapes (forests and other greenery) and memories of gardens and gardening are commonly invoked by gardeners. These memories propel residents to join community gardens and devote time and effort to the creation of communal space. The image and experience of

---

1   Pile, S. 1996. *The Body and the City: Psychoanalysis, Space and Subjectivity*. New York: Routledge, p. 12.

the garden symbolizes for many gardeners their pre-urban life, and in some cases their pre-United States life—as the story of Ilya, a gardener in the Bronx, suggests:

> I'm from Iran ... so I think that was a big part of my connection to nature. Then in Shiraz, where I grew up—that is the city of gardens and poets—the whole city is like trees and plants and parks and gardens in the middle of the house as well as outside and lots of fruit trees; those were really highly praised. We always go out and pick apples or we go out and get some walnuts and get all our hands black, or we go out and take some berries and mulberries ... so it was always a big part of our growing up ... to be part of climbing and eating and playing and natural stuff.[2]

The symbolic reconstruction of past landscapes in the space of the garden is also accompanied by the reenactment of past practices and experiences of space. Most stories of urban gardeners included here, like that of Ilya from Iran, include a variety of practices that used to be part of their everyday life in the past and that are afforded again by these gardens. Sam, a gardener from Harlem, Manhattan describes his experience in a similar tone:

> I come from New Jersey, which is the garden state and I worked on farms and stuff like that in the past; and when I was young, when I was growing up, my parents always had a garden in the backyard and we had a grapevine and peach tree, and we grew collard and pepper and tomatoes and stuff like that; and my mother did a lot of canning of vegetables and fruits and stuff. Because they came from the South, you know, so what was really important for them was having a stable community, taking the crops and storing them. So we become self-sufficient. So they passed some of that along to me and I became involved with the garden. So we planted the cherry tree, palm trees, mulberry trees, the flowers, and vegetables.[3]

Not only does the reconstructed landscape help gardeners familiarize themselves with the space of the city, it also allows for meaningful engagement with the space by enabling them to practice a part of their past repertoire of behaviors. Picking fruits, growing fresh produce, gardening, canning, and being self-sufficient are practices that the city, generally speaking, does not support in either private or public forms. In the gardens, however, residents establish a space where these practices are not only being supported but also celebrated.

The rural ethos that dominated the American geographical imagination for generations is an important factor in the powerful experience of a green oasis in the midst of a massive grey urbanity. It is an ethos that reflects either gardeners' life experience or a narrative communicated to them by parents and wider family.

---

2   Interview, May 31, 2005, South Bronx.
3   Interview, June 8, 2005, Harlem, Manhattan.

The perceived superiority of the rural environment over the urban shapes the needs and attitudes of urban residents. Expressions such as "I missed gardening because I grew up gardening and New York City was hard to me because it is so urban and I grew up in a more rural environment,"[4] voiced by Claudia—a gardener from East Harlem (and originally from New Mexico)—reiterate that the gardens can address needs by offering the opportunity to (symbolically) reconstruct past landscapes and re-enact past practices.

For the gardeners, the garden is a place where a sense of familiarity and belonging can flourish alongside the process of assimilation to life in the city. Juxtaposed to an urban environment dominated by Western principles of design, the garden makes the surrounding space more familiar. In this space, and because of it, gardeners can relate to their neighborhoods and more fully recognize themselves in them. Moreover, the gardens persist as a "balancing factor" that moderates the excessiveness of the urban experience by symbolically representing a remembered, more familiar, and longed-for open, wild, natural landscape. As Ilya from Iran suggests: "And again, oxygen. I wouldn't live in New York City if there wasn't a place I could put my hands in dirt, and play, have fun, and make a living out of that and be able to laugh and enjoy myself."[5]

Many gardeners, in particular those who are parents, view the garden as a connective tissue not only between past and present but also between present and future. They stress their wish to offer their children something that was significant to them in their childhood as well as the special experience that was afforded to them by the garden. A representative from the Council on the Environment of New York City (CENYC) makes the direct connection between his experience and its implication for his children:

> I grew up in Queens but I was in a single house. We had a backyard that my father used to grow vegetables and I had to go out and pull weeds but I was also able to pick the radish right out of the ground and the string beans right off the plant and eat them fresh, and cherry tomatoes, cucumbers. So I think that having that experience definitely is a positive thing. And one other thing, I have two small children and they know about nature and they know about gardens, obviously, not because I am forcing it on them but [to] make sure they know about these things.[6]

Sometimes the gardens represent not a reconstruction of past landscapes but actually their negation; but even then they are clearly integrated into the environmental autobiographies of the gardeners and constituted as a completion of and compensation for something that was missing in their lives. The story of

---

4   Ibid.
5   Interview, May 31, 2005, South Bronx.
6   Interview, December 5, 2005, Manhattan.

Mr Thomas, a gardener from East New York who grew up in one of the housing projects of Upper Manhattan, exemplifies this point:

> And slowly I began to learn a little more because I'm not a gardener at heart. I don't know anything about this. I'm concrete. I'm the city guy. But I do love fresh fruits and vegetables and I love the earth I love the greenery I love the quiet … like what we are doing now [sitting at the garden]. I can do it forever. And I began to learn from other gardeners.[7]

Some spatial features of the garden invoke more liminal memories, ones which were not personally experienced. The gardens connect the present manmade environment with a pre-modern one—wild, pristine—by making explicit that the place had a past of its own, one which was dramatically different from the current and which is independent of human action. Mike shrewdly perceives this connection in his garden in the Lower East Side of Manhattan:

> [The garden] is an indicator of what it used to be [like] here. The willows show that it is a wet area; it is a constant reminder that there are creeks and [a] tidal estuary, [a] huge tidal estuary. I would like to make that more obvious by maybe daylighting the creek.[8]

Community gardens also function as a site of social history—the memory of the origin. The space, alongside photos and stories of gardeners, serve as a mnemonic for the bad conditions of the site and the neighborhood before the gardens were established. In addition, many gardens are used as memorial sites to commemorate loved ones or important figures from the neighborhood or from gardeners' history and culture. These gardens are usually named after the person being remembered and present commemorative objects such as sculptures, murals, photos, and signs.[9] Using the space in these ways reinforces *civic memory*—"the recollection of the events, characters and developments that make up the history of one's city or town." Civic memory is a crucial component that draws residents together or, in the case of community gardens, members, and "generate[s] a sense of civic identity."[10]

Experiencing past landscapes as an integrated component of the everyday experience of urbanity invokes and complicates the awareness of residents with

---

7   Interview, May 6, 2006, East New York, Brooklyn.

8   Interview, October 5, 2006, Manhattan.

9   Martinez, M. 2002. "The Struggle for the Gardens: Puerto Ricans, Redevelopment, and the Negotiation of Difference in a Changing Community." Unpublished dissertation, New York University.

10   Dagger, R. 2000. "Metropolis, Memory and Citizenship." In Isin, E. (ed.). *Democracy, Citizenship and the Global City: Governance and Change in the Global Era.* London: Routledge, pp. 25–47, p. 37.

regard to their space and environment. The significant component here is spatial diversity. The "everydayness" of their interaction allows gardeners to retain two or more (sometimes contrasting) mental images and sets of attitudes and effects with regard to their environment. As a result, both dominant and alternative environments (with their complexities) become more salient. The idea that an alternative experience of the environment is possible undermines the hegemonic image of the city. The fact that gardens are their own product facilitates an understanding that they themselves made the alternative happen. This important realization makes issues of space and environment even more prominent in gardeners' consciousness.

Aesthetic Experience and Life as Celebration

Aesthetics and celebration are central to people's experiences of gardens, but they are usually given little attention in our understanding of the experience of space (and urban space in particular). The reason for the inadequate treatment that these two subjects are given in social sciences can be attributed to their decreased importance in modern life, governed by instrumental rationality. In fact Lefebvre, in his seminal critique of everyday life, perceives the two as essential enablers for the process of de-alienation of the everyday life.[11] Aesthetics and celebration are integral components in the interaction of urban residents with their environment. In the context of New York City community gardens aesthetics and celebrations contribute both to individuals' everyday experience of the city and, as shown later, serve as a strategy of struggle over space, as a means for its production and as a collective resource for power and knowledge (described in Part III).

Aesthetics, in this case, refers to a sensory experience of the environment which is of beauty and satisfaction, as well as to the opportunity to be involved in creative activities that are perceived to yield aesthetically pleasing results. A variety of traits are considered as rendering cities "aesthetically pleasing":

> [E]thnic and cultural variety [...] a diversity of vegetation, [...] public art and freedom of expression in the community in the forms of sculpture, graffiti and street art, a range of build-out (or zoning) that creates both densely and sparsely populated areas, scenic neighboring geography (oceans or mountains), public spaces and events such as parks and parades, musical variety through local radio and street musicians, and enforcement of laws that abate noise, crime, and pollution.[12]

---

11   Lefebvre, H. 2002. *Critique of Everyday Life: Foundations for a Sociology of the Everyday* (vol. 2). New York: Verso.

12   According to Wikipedia, retrieved July 19, 2007: http://en.wikipedia.org/wiki/Aesthetics#Urban_life.

Scholars from various disciplines echo this list of elements.[13] Community gardens in New York City allow each of these traits to materialize in various degrees, and therefore to contribute to the aesthetic experience of gardeners and urbanites in general.

The gardens provide a unique aesthetic experience which differs from the rest of the built environment and from urban parks. They all differ in vegetation (trees, bushes, flowers, fruits, vegetables, water plants, herbs, etc.); animals (mostly birds, butterflies and bees, turtles, fish, squirrels, hens, etc.); structures (gazebos, benches, tool sheds/casitas, amphitheaters, stages, etc.); design elements (individual/ common plots, elevated plots, lawns, ponds, etc.); art displays (sculptures, murals, mandalas, etc.); and art events (theater and music performances, movie screenings, art exhibitions, etc.). Visitors to the gardens can therefore engage with a wide variety of assemblages of colors, light, levels of seclusion, smells, sounds (birds singing, leaves in the wind, music, etc.), and sensations (cool, breezy, shady, etc.) according to their preference.

Figure 2.1   "Personalizers" in individual garden plots

---

13   See for example Botkin, D.B. and C.E. Beveridge. 1997. "Cities as Environments." *Urban Ecosystems* 1(1): 3–19; Mattila, H. 2002. "Aesthetic Justice and Urban Planning: Who Ought to Have the Right to Design Cities?" *GeoJournal* 58(2–3): 131–8.

Gardeners are involved in the creation of their own aesthetic vision through the design of the landscape—assembling flowers, plants, stones, and structures. It is very common to find additional decorations such as stones and sculptures in gardens that personalize individual plots (see Figure 2.1). The gardens exist as open creative outlets in the city for both observers and producers of art. They are places where people can exhibit or perform their art regardless of their proficiency. Finally, the gardens are some of the only places in the city where public art—a movement that flourished in the 1980s and had subsequently been eradicated from the city's streets[14]—can still be encountered. An interesting example is the many murals that cover the buildings on the perimeters of the gardens that express various social and political commentaries: some deal with the struggle for land that was waged by gardeners, others with abstract issues such as freedom, community, nature, etc (see Figure 2.2).

Figure 2.2   Murals

This richness of aesthetic experience, being less prevalent in urban life but abundant in gardens, is not taken for granted by the gardeners. The words of Ilya, from the Bronx, poetically capture the unique nature of the gardens:

14   Candela, I. 2007. *Sombras de Ciudad: Arte y Transformacion Urban en Nueva York, 1970–1990.* Madrid: Alianza Editorial [in Spanish].

> Garden[ing] is probably the perfect art project because it works with the
> community; it lives by itself. I always loved art projects that when I made, started
> creating their own life. Other people work with them, or an installation; somebody
> do this and that and it started do its own thing but garden[ing] does that naturally—
> it grows, people plant thing, we have sculptures, we got performances, kids grow
> up in it. And so it's an ideal and it's all the people who are the community sharing
> their vision of what they believe and what they want.[15]

The idea of celebration interacts with that of aesthetic experience. It involves
being and doing for the sake of being and doing, or for the sake of joy, outside the
realm of instrumental rationality. Celebration as an activity in the garden is usually
manifested in parties, festivals of various kinds, street parades, neighborhood
cookouts and feed-ins, music and dancing, and various leisure workshops and
programs for children and adults alike. But celebration is also manifested in the
simple, less orchestrated activities such as gardening, socializing, playtime with
ones' children, finding a place of serenity, observing nature and the joyful activity
of others, and reading. In short, having a place where the practice of leisure is
abundant and encouraged. As Claudia, a gardener from East Harlem observed:

> I think people are feeling connected to nature and it relaxes them to be in the
> garden. I think a lot of gardeners, a lot of them do, but some of them don't
> necessarily know much about gardening but do it because they enjoy being in
> the space and getting together.[16]

Emily, a gardener from the East Village, provides a taste of this richness through
her own experiences:

> We attend most of the workshops and musical events and we work, you know,
> to help on the events that they have. I coordinated the last Christmas party [...]
> I am very involved in that way and that is really great for me. I mean it is great
> for me to be able to go out with my daughter to see some excellent music for
> free in the evening right across the street from my house. It is very easy. There
> were different workshops we have attended: we attended a fish printing workshop
> where [my daughter] printed fish on clothing; we have attended the mandala
> making workshops when they were making the fence and tracing the hands 'cause
> these are all garden members' hands, and the hands of the children there. In fact,
> my daughter's hands they were very little at the time [...] We have attended
> just numerous workshops, tie-dyed cloth, making jewelries, and then I taught
> workshops in marble games, making rubber bands balls, and making baskets.[17]

---

15   Interview May 31, 2005, South Bronx.
16   Interview, November 8, 2005, East Harlem, Manhattan.
17   Interview, April 24, 2002, East Village, Manhattan.

Emily's story raises several important issues: it describes the variety of the experiences possible in community gardens; it highlights the importance of all this abundance being free; and finally it underlines the opportunity to (re)create oneself in space through celebration.

Beyond the openness of access that allows all people—the haves and the have-nots—to engage in celebration and aesthetic experience, cost-free activities lend themselves to an increased freedom of engagement. The regular rules of transaction between participants are more flexible and allow a wider spectrum of forms of participation for both performers and audience. As a result, activities are less formal and less calculated, and participants can therefore take a more active part in a performance or a workshop than they could ever have in paid-for activities. In addition, these cost-free opportunities for celebration are anchored in their relative scarcity rather than their multiplicity, that is, the divergence of having-so-much-for-free from life experiences outside the garden. Where else do people have the chance to experience such wealth? What happens in the garden is not charity; the roles of givers and receivers constantly interchange and there is a wealth of shared resources in the artistic and social skills of the people and in the physical setting. The discrepancy between the common wealth of the gardens and the lack thereof in other realms of life is intriguing for gardeners and other participants. They are aware of this discrepancy, which not only recharges the value of the garden but also leaves unsettled the places and activities outside of the common wealth. This unsettling experience of the world and awareness of existing discrepancies is one of the "concientizing" effects of the gardens. It is a step towards mitigating *prescription*—a behavior of the less powerful that is prescribed by dominant guidelines.[18] The experience of an alternative challenges the prescribed dominant experience. Concientization, which entails mitigating prescription, is the process of rethinking and rearticulating a world view through a critical gaze.[19]

Back to Emily's story and other important issues it raises: such as the issues pertaining to collective creative abilities and (re)creating oneself in space. These issues are presented through the story of the mandala that the garden members produced.[20] Mandala-making interweaves the aesthetic and the social in "a unifying experience in which people can express themselves individually within a unified structure."[21] Emily can find the traces of her own history as it is engraved in space by her daughter's small palms on the mandala fence (see Figure 2.3). These objects that are imprinted with the people who created them—sometimes literally, like the fence, and sometimes more figuratively—serve as social mnemonics not only for their creators but also for others who encounter them.

---

18 Freire, P. 1972. *The Pedagogy of the Oppressed*. New York: Penguin.

19 Freire, P. 1971. "A Few Notions about the Word 'Concientization.'" *Hard Cheese* 1: 23–8.

20 "Mandala" refers to the process of collective creation of an art object.

21 http://www.mandalaproject.org/What/Index.html.

The collective creative activities that often occur in community gardens are instances of communal determination of space. This process of production may be understood as what Marx termed the *objectification* of species-being, "for they duplicate themselves not only in consciousness, but actually in reality."[22] It is through objectification resulting from collective experience and action (or, for Marx, "positive labor") that people can "contemplate themselves in the world they have created" and at the same time become "object[s] for others within the structure of social relation[s] and in this way create civilization."[23] Objectification reverses species alienation, the separation and atomization of human consciousness and experience that Marx saw as the consequences of dehumanizing labor that characterizes capitalist societies. Objectification therefore is a means towards de-alienation and emancipation.

Figure 2.3   The mandala at 6&B community garden, Manhattan

Ehrenreich's (2006) account of the demise of ecstasy and joy in European and American cultures suggests that the "emerging capitalist perspective [with its

22   Marx, K. 1959. *Economic and Philosophic Manuscripts of 1884*. Moscow: Foreign Languages Publication, p. 114.

23   Morrison, K. 2006. *Marx, Durkheim, Weber: Formations of Modern Social Thought*. Thousand Oaks: Sage, p. 404.

relentless] focus on the bottom line" presents a growing desire for "well-regulated human labor." In this way leisure came to represent "the waste of a valuable resource [... as] festivities had no redeeming qualities. They were just another bad habit the lower classes would have to be weaned from [...] recuperate from, the weekend's fun."[24] Ehrenreich uses Turner's notion of *communitas* as a useful tool for her account.[25] Turner "recognized collective ecstasy as a universal capacity and saw it as an expression of what he called communitas, meaning, roughly, the spontaneous love and solidarity that can arise within a community of equals."[26] However, Ehrenreich attaches a different social function to communitas than Turner, who believed that communitas should be marginalized and allowed as very occasional relief that will keep the social structure from becoming overly rigid. Ehrenreich on the other hand stresses the importance of this repressed capacity for the generation of inclusiveness, social cohesion, happiness, and integrity. She suggests that "the urge to transform one's appearance, to dance outdoors, to mock the powerful and embrace perfect strangers is not easy to suppress. [...] The capacity for collective joy is encoded into us [...]. We can live without it, as most of us do, but only at the risk of succumbing to the solitary nightmare of depression."[27]

Celebration and aesthetics, therefore, in many ways are left outside of the hegemonic order of everyday life. Together, according to Lefebvre (following Marx's vision for the transformation of everyday life), they are key to the process of transformation that may result in the de-alienation of everyday life. Lefebvre (2002) refers to the projects of transformation that rely on aesthetics and celebration as "*aesthetic* in nature." It is a level of being of "higher creative activity" in which enjoyment of the world is not "limited to the consumption of material goods" but driven from the rediscovery "of the spontaneity of natural life and its initial creative drive, [... and] art would be reabsorbed into an everyday." According to this vision everyone would "perceive the world through the eyes of an artist, enjoy the sensuous through the eyes of a painter, the ears of a musician and the language of a poet. Once superseded, art would be reabsorbed into an everyday which has been metamorphosed by its fusion with what had hitherto been kept external to it." Thus, the spiritual powers of humans which are now alienated will make the "journey back to ordinary life and invest themselves in it by transforming it."[28]

---

24   Ehrenreich, B. 2006. *Dancing in the Streets: A History of Collective Joy.* New York: Metropolitan Books, pp. 100–1.
25   Turner (1969) in Ehrenreich, 2006.
26   Ehrenreich 2006, p. 10.
27   Ibid. p. 260.
28   Lefebvre 2002, pp. 36–7.

The Question of Identity

> The formation of self, self-awareness, identity and so on is a geographical matter
> since "people are socialized in localized contexts."[29]

It is suggestive that the simple question of why one is a gardener evoked respondents' environmental autobiography. Answers to such a question could have been practical: the desire (or need) for fresh and tasty tomatoes and cucumbers; the simple pleasure of gardening as a leisure activity; the need to belong to a community—most gardeners would agree with such answers. But gardeners go beyond the practical level and make the connection between their past landscape and practices and their experiences and activities in the garden. The garden becomes integrated into their life stories, their environmental autobiographies, in a way that maintains its continuation and the continuation of their "self." Being a gardener, with all that it entails, serves as a validation of that self.

As a direct product of gardeners themselves, community gardens not only persist as spaces that resonate with the past experiences of gardeners but also constitute and emphasize the gardeners as the "producers of space." There is a strong sense of connection that is gained by all the work of building, designing, gardening, socializing, and fighting for the garden. As Claudia conceptualizes her own connection to the garden:

> The time that you sweat and worked in that place that is what makes you feel
> connected to it. And gardening is definitely a labor of love that takes years and
> a lot of patience. And when you've done that and you put that time into it over a
> long time, now you feel connected to it.[30]

Being part of a garden, which in most cases implies being the producer of space,[31] lends a sense of control and ownership over the environment. These relational elements with the environment facilitate a multi-layered experience of it, and through this rich experience gardeners increasingly identify with the environments they made. In turn, this process of identification reinforces the environment's "symbolic significance as a substrate of social, emotional and action-related contents." Through this dynamic and dialectical process the environment becomes "subjectively meaningful."[32] The process of identification with the environment that gardeners go through in their

---

29   Thrift (1989) in Pile, S. 1996. *The Body and the City: Psychoanalysis, Space and Subjectivity*. New York: Routledge, p. 68.

30   Interview, November 8, 2005, Manhattan.

31   There are several cases (specifically among NYRP's gardens) in which the role of the gardeners as producers was downplayed. Chapter 3 discusses the implications of this change on the relations between individuals and space.

32   Lulli, M. 2001. "Urban-Related Identity: Theory, Measurement, and Empirical Findings." *Journal of Environmental Psychology* 12: 285–303, p. 285.

interaction with the garden results in the construction of what may be called place identity,[33] urban identity,[34] or sometimes green identity[35]—which are sub-structures of the psychological concept of self identity.[36] Connecting past and present gardens helps sustain a continuous sense of self and space and supports the recognition function of place identity. As a place in which gardeners can exercise a high level of control and that serve as an outlet for self-expression and creativity, the garden provides a sense of security and meaning.[37]

The degree to which space feeds into gardeners' self-perception is evident from the many cases where the garden is given a propulsive power in the progression of gardeners' life stories. The strong integration of the garden into their story endowed the garden with a constitutive role which has an almost mythical stance in their personal development beyond the garden per se. Mike, a gardener from the East Village, Manhattan, talks about the important part the garden played in his career as a landscape designer:

> So indirectly [the garden] actually helped me build my career and start my business [...] I was working for C. and getting all these high-end clients. And at the time, and we still do, we get all these extra plants from jobs. We turn over a garden and all the plants that we have left over we bring them here [to the garden]. So we were starting to get this huge collection of different, weird groupings of plants. So this was coming together, my career was coming together, it was very symbiotic.[38]

Later on this gardener goes on to discuss the role that the garden played in bringing him and his life partner together.

Beyond introducing spatial diversity and a readily available space for identification with the environment, community gardens also afford human diversity. A wide variety of people can find their place in the garden. This diversity tends to produce internal conflicts, or what gardeners generally define as "personality issues" that sometimes aggravate animosities. But in most cases gardeners just learn to work together despite their differences. Although originally established and managed by predominantly African American and Hispanic people of low socio-economic status, urban transformations make the gardens a unique

---

33 Proshansky, H., A. Fabian, and R. Kaminoff. 1983. "Place-Identity: Physical World Socialization of the Self." *Journal of Environmental Psychology* 3: 57–83.

34 Lulli 2001.

35 Horton, D. 2003. "Green Distinctions: The Performance of Identity among Environmental Activists." In S. Bronislaw, H. Wallace, and C. Waterton (eds), Nature Performed: Environment, Culture and Performance. Oxford: Blackwell, pp. 63–78.

36 Proshansky, Fabian, and Kaminoff 1983.

37 Ibid. The main functions of place identity are recognition function, meaning function, expressive-requirement function, and mediating change function.

38 Interview, October 5, 2006, East Village, Manhattan.

environment where people of different class, ethnicity, living conditions, and life aspirations can find their place and work in collaboration with each other as well as establish personal connections.

As discussed in Chapter 1, community gardens are clustered in poorer neighborhoods of underprivileged residents. However, with the urban restructuring of the 1980s other types of residents, from higher socio-economic status, were introduced to the gardens. Neighborhoods awash with community gardens—such as the East Village and Harlem in Manhattan and Bedford-Stuyvesant (and others) in Brooklyn—were changed, gentrified, and now accommodate a greater range of people: people living on government subsidies work together with the better-off (newer) residents.

The gardens accommodate a variety of needs and allow for multiple outlets. Most gardeners talk about their inner psychological needs that the garden addresses, some more explicitly than others. Joshua, a gardener from Manhattan, searched for a sense of connection and love and a sense of achievement and competence. He describes his reasons for becoming a gardener in this particular garden:

> I wanted to nurture something and I figured plants are easier, they don't talk back. So that was what initially got me started. And one thing on my initial tour of this place […] I noticed the pond and I thought I liked the pond, I didn't know if there were any fish in it, it looked like it was kind of falling in and I just kind of made it my project.[39]

A representative of Green Thumb provides a broader understanding of the gardens in the city and the human mix they nourish:

> There are a lot of people who are gardeners who are sort of strange. How do you say that in a nice way? But I know of some gardeners who are seriously mentally ill that can function in the garden and you can't imagine them functioning in society. It is a way for them to interact with other people in a space that they have some kind of control […].[40]

Mike, from the East Village, Manhattan, reflects on the same subject, fine-tuning and personalizing the previous observation:

> I think that for someone like S., and she is coming here for years, she was part of the original legal battle. Like it or not but she was very helpful when she was more lucid and she is also very smart and this is very important thing for her. A lot of people they really need it. And even if the personality issue can be really difficult sometimes, I value pretty much all the members and I think that you kind of see, as soon as you notice some of the difficult personalities

---

39 Interview, September 27, 2006, East Village, Manhattan.
40 Interview, November 15, 2006.

you realize that this is why they are here, they need this, they need an outlet and they need to be heard, they need to take care of something, to have some saying in something that matters. I'm trying not to sound over-dramatic or over-Pollyanna, and I'm certainly one of those persons. I certainly needed something when I came here and I got it. I needed to belong to be productive and to be valued.[41]

These comments are indeed very insightful as to the found opportunity for self-realization and gratification that various types of people can find in community gardens.

The changing character of the neighborhoods and, in turn, of gardeners also brought about difficulties and conflicts that in some gardens resulted in increasing exclusionary attitudes rather than harmonious acceptation of social diversity. Gardens of the East Village, for example, used to be dominated by Puerto Ricans until the mid-1980s when newcomers, "artist gardeners," arrived in the neighborhood. The neighborhood in the first decade of the twenty-first century presents a greater ethnic diversity (relative to the Hispanic enclave it was before the 1980s), but the gardens in it (about 30) maintain the predominance of either white or Latino gardeners. Therefore, the ethnic lines were not erased and actually might have been reinforced.[42]

A somewhat similar process of gentrification recently began in Bedford-Stuyvesant, Brooklyn. The neighborhood, which as of the 2000 census was 77 percent African American and 18 percent Hispanic, experienced a dramatic appreciation in its housing value, which was accompanied by changes in its ethnic and socio-economic structure. The implications for community gardens in the neighborhood (over 50 of them) are new negotiations over the purpose and essence of the gardens. "It is totally a different world," said Edie Stone, executive director of Green Thumb, of the situation in Bedford-Stuyvesant:

It is not a bunch of people from down south growing collard greens and vegetables anymore. It is like they [the newcomers] want to have a stage and they want to do programming and they want to have kids' projects and all these kind of things that more affluent families want to do. Which there is nothing wrong with that, it is just different. When the gardens were interim it was just like so much people wanting to fill in the space; they wanted to make it better, they didn't want it to be bad. And now it is the opposite. The spaces are premium and people are fighting over what you can do with these spaces.[43]

---

41   Interview, October 5, 2006, East Village, Manhattan.
42   Martinez, M. 2002. "The Struggle for the Gardens: Puerto Ricans, Redevelopment, and the Negotiation of Difference in a Changing Community." Unpublished dissertation, New York University.
43   Interview, November 15, 2006.

East New York, Brooklyn, that was home to an 87 percent African American and Hispanic population in 2000[44] has since experienced waves of immigrants from Pakistan that do not necessarily change the socio-economic structure of the neighborhood but change the neighborhood cultural structure. Conflict between the veterans and newcomers were so intense that Green Thumb had to intervene in order to resolve the tension. Edie Stone found in the neighborhood:

> a lot of strange situations where a whole bunch of Pakistani immigrants move in and they have an agricultural tradition and they are totally interested in gardening, but the African Americans who were there before totally don't trust them and find them very strange. And they['re] like, "I don't know, they don't do things the way we like to do" and we have to be like, "Well you can't tell anybody they can't join, you just have to figure out how to deal with that." And that is one reason why I was doing so many conflict resolution trainings.[45]

Since this is not a unique situation Green Thumb offers workshops in conflict resolution and gardens can invite in conflict resolution counselors in order to deal with a specific conflict.

Changes in the structure of neighborhoods influence the social dynamic within the gardens and generate some challenges, but at the same time they bring to the surface negotiations over the spirit and meaning of gardens: gardens as a neighborhood resource for fresh produce and fresh air; gardens as a place for cultural expression; gardens as a recreational facility or as a site for exercising ideologies (for example, environmental stewardship).[46] Despite internal challenges, the gardens bring together individuals from all walks of life and maintain social diversity by allowing for different types of activities and different engagements with place to occur. The gardens offer a setting for coincidental interactions between gardeners and non-member visitors. Moreover, their daily operation requires that gardeners collaborate in meetings, working days, and the planning of events. Sam from Harlem describes the role of his garden in the neighborhood:

> [The garden is] important as a place of meeting. We have our block association meetings in summertime out in the garden and people in the neighborhood want

---

44 2000 census information.

45 Interview, November 15, 2006.

46 Zukin, S. 2010. *Naked City: The Death and Life of Authentic Urban Places.* Oxford: Oxford University Press. Zukin points to a certain sequence in which the meaning of urban practices (what she understands as authenticity) changes. According to Zukin, the meaning of urban community gardening changed from a grassroots movement that contests the state, to represent ethnic identity, to secular culture, and finally to sustainability ideals of urban food production. Each of these forms is the outcome of a different ethnic group and social class arriving at the gardens (Zukin 2010, p. 197). However, different meanings, ideologies, and authenticities can be found concurrently and feed each other in gardens across the city.

to come so we gave them keys to the garden so they can go in at any time and use the garden. It became a meeting place, community center so to speak where people can come together, have cookouts and just sit and enjoy, read a book.[47]

The gardens also offer an informal mechanism to get involved in the community and become knowledgeable about it, and to cultivate a local social network. Joshua from the East Village explains:

I live a little bit further, to the west of here, so I'm not as plugged to the dynamic of the little changes that are going on up here but I love this neighborhood so it [the garden] is a good way to get to know around. And it is also the social issue too. I figure I could meet people here with common interests and I made friends.[48]

Mike, another gardener from the East Village, emphasizes the role of the garden for social interaction and the opportunity to establish social relations within the living environment. He compares the garden with the nearby public park and suggests that:

We tend to be really introverted in our life style now because we deal with computers we don't deal with people so much anymore on a regular basis. I didn't know anyone in my building when I got involved in the garden. I certainly didn't know anybody on the block. So there is that, but also having this public arena for performance and gathering I think it's incredibly valuable and it doesn't happen anywhere else. It doesn't quite happen in Tompkins Square where you could just have an informal or very formal gathering in the space where very little money [is] put into it, very little effort.[49]

The gardens assist residents in familiarizing themselves with their social environment and getting to know their neighbors. At the same time they not only function as another neighborhood's stoop (or park bench) but also provide opportunities to experience different types of interactions that involve working together, enjoying art and nature together, and creating something new together.

The chaotic, degraded, unsafe, and unpleasant environments that residents faced throughout and after the fiscal crisis of the 1970s could not afford a mutual reinforcement between the self and the environment. On the contrary, if anything such environments cause avoidance, withdrawal, and alienation.[50] Moreover, modern urban life in general, as the last gardener explicitly suggested, tends to

47   Interview, June 8, 2005, Harlem, Manhattan.

48   Interview, September 27, 2006, East Village, Manhattan.

49   Interview, October 5, 2006, East Village, Manhattan.

50   See for example Latkin C. and A. Curry. 2003. "Stressful Neighborhoods and Depression: A Prospective Study of the Impact of Neighborhood Disorder." *Journal of Health and Social Behavior* 44(1): 34–44; Nasar J. and B. Fisher 1993. "'Hot Spots' of

alienate people from each other and from their environments. Nevertheless, by engaging with their surrounding space and producing gardens as environments that afford meaningful and supportive relations, it became possible for residents to alter this alienation. The gardens became sites of strong identification and function as significant components in the constitution of positive place identity among gardeners. Since community gardens beautify neighborhoods and contribute to their safety, they create the conditions for identification with the neighborhood at large and contribute to positive place identity among non-gardening residents as well.

Construction of Identity in Relation to Space

In the era that the "death of the subject" was announced, when the concept of identity is challenged and its common understanding as a consistent objectification of our selves is shuttered,[51] the discussion of identity is much more tortuous. Alternatively, we are advised to understand the "subject" as an "assemblage that metamorphoses or changes [its] properties as [it] expands [its] connections" in regard to other objects, practices, and forces.[52] Thus, awareness of the external world, social and physical, constitutes one's being and in a way "imposes" one's identity. The contemporary discourse of identity is, then, that of hybrids and multiplicities. Yet we still understand these through the narratives (multiple narratives) that are being attached to the selves. We have seen that the experience of the garden composes important parts of the life stories of the people involved. This strong connection to the place makes it a significant component in the assemblage of the subject. Since it is an "assemblage that metamorphoses," it is only possible to speak of the process of assembling the self (rather than the completed product). Therefore, we seek to examine only the components that play a significant role in this process.

The opportunity for social relations within one's residential setting along with satisfaction with the environment, the level of choice that can be exercised in it, and a sense of ownership are the essential factors that enhance the process of identification.[53] Community gardens, the argument goes, embody a multiplicity of the essential conditions for identification with the environment and the development of a prominent place identity . At the same time this multiplicity of conditions caters to a highly diverse group of people. In this way that space of the garden is capable of embracing various identities while reproducing them as multiplicity; it can both contain and reproduce social diversity.

Fear and Crime: A Multi-Method Investigation." *Journal of Environmental Psychology* 13: 187–206.

    51   Rose, N. 1998. *Inventing Our Selves: Psychology, Power, and Personhood*. Cambridge: Cambridge University Press.

    52   Ibid., p. 172.

    53   Lulli 2001; Proshansky, Fabian, and Kaminoff 1983.

The idea that the space of the garden and the practices that it establishes allows for diverse identities to come together, collaborate, and get a unique experiential value out of it rests on the welding of everyday life activities and the space in which they occur.

A community can be understood as "a space of *emotional relationships* through which *individual identities* are constructed through their bond to *micro-cultures* of values and meanings."[54] Identity, then, can be seen as the product of the physical setting and social relations of the everyday life; or, in other words, identity is the product of a community. Psychological identification with the community proposes a relation that appears less "remote" and more "direct," one which occurs not in the "artificial" political space of society but in matrices of affinity that appear more natural. One's communities are nothing more—or less—than those networks of allegiance with which one identifies existentially, traditionally, emotionally, or spontaneously, seemingly above and beyond any calculated assessment of self-interest.[55]

A community necessitates social practice that is based on cooperation and collaboration "unmediated by money and the value practice of capital."[56] It still remains to show the ways in which the community that is formed in community gardens constructs political identities or subjectivities as it is being reinvented by conscious political action.

Three facets of community gardens are the foundation for molding the living environment in the image of its inhabitants. The first facet is related to the molding of space to resonate with the familiar landscapes of the past, childhood landscapes, landscapes of competence, landscapes of love and longing. The second facet encompasses aesthetic and celebration experiences. It captures how participants are able to reintegrate collective festivity and artistic and cultural practices into their life through their gardens. These practices—increasingly eliminated from the public sphere as their existing manifestations are commodified—have spontaneous and unfettered expression in the garden. The third facet is the construction of identity in relation to space. The experience of space in community gardens as inclusive and supportive of the needs of gardeners and as an expressive space of their investment facilitates gardeners' identification with the space, which in turn supports the formation of their identity. In the same vein the gardens are used as a site to struggle for and express the identity of urban residents.

---

54 Rose, N. 1999. *Powers of Freedom: Reframing Political Thought*. Cambridge University Press, p. 172 (emphasis in the original).

55 Ibid., p. 177.

56 De Angelis, M. 2007. *The Beginning of History: Value Struggles and Global Capital*. London: Pluto Press, p. 66.

Chapter 3
Determining Space, Creating Opportunities

Sense of Ownership and Control over Space

Participation and control over space are strongly intertwined. Participation—the direct involvement of residents in designing, constructing, or managing their living environment—is a tool to achieve greater sense of control. Sense of control is important for the development of attachment to, identification with, and meaning of place, as well as the development of psychological ownership—all of which have direct psychological, behavioral, and social outcomes.[1] In general, residents' control over their living environment challenges the growing sameness of the urban by "make[ing] favorable differences in the lives of the public."[2]

The concept of psychological ownership is particularly useful for examining affects towards a communal or public space. It generally refers to the "feeling of possession in the absence of any formal or legal claims of ownership."[3] It is a psychological structure that has an objective stance in reality[4] which, unlike legal ownership, is recognized by the individual rather than by society.[5] Psychological ownership entails the experience of close connection with the object of possession to the effect that it is perceived to be part of the extended self. Since psychological ownership usually develops out of feelings of responsibility and concern for the object of possession,[6] its development, especially towards community space, is a function of the level of participation.

Gardeners frequently talk about their affection for their garden and express this affection in various forms. A strong sense of connection is gained from all the work of building, designing, gardening, socializing, and struggling for the garden. These activities of producing the garden facilitate a sense of belonging

---

1   Francis, M. 1989. "Control as a Dimension of Public-Space Quality." In I. Altman and E. Zube (eds). *Public Spaces and Places*. New York: Plenum.

2   Ibid., p. 154.

3   Mayhew, M., N. Ashkanasy, T. Bramble, and J. Gardner. 2007. "A Study of the Antecedents and Consequences of Psychological Ownership in Organizational Settings." *Journal of Social Psychology* 147: 477–500, p. 477.

4   Etzioni, A. 1991. "The Socio-Economics of Property." *Journal of Social Behavior and Personality* 6(6): 465–8.

5   Pierce, J., M. O'Driscoll, and A. Coghlan. 2004. "Work Environment Structure and Psychological Ownership: The Mediating Effect of Control." *Journal of Social Psychology* 144: 507–34.

6   Parker, S., T. Wall, and P. Jackson. 1997. "'That's Not My Job': Developing Flexible Employee Work Orientations." *Academy of Management Journal* 40: 899–929.

and attachment to place. In addition, the high level of control that gardeners can exercise over space evokes a sense of ownership, or psychological ownership towards the land and the neighborhood. As gardeners have no legal title to the land,[7] neither have they produced an exchange-value of it, the concept of psychological ownership plays an important role in the attitude towards and practices of residents in communal or public spaces.

People generally have a perception of "ownership" that contradicts the perception of "public" or "communal." Claudia, a gardener from Harlem, delves into this contradiction:

> […] how and why people feel connected to land, to space which is not their
> own private property? I don't think there are a lot of such occasions that you
> can observe that in our society, in our culture. For the most part our connection
> to land has to do with property ownership. Not shared use of a common space.
> […] I definitely think that the feeling of connection to the land and that kind of
> ownership—I'm describing it as ownership even though it is not an ownership,
> it is funny that this is the context I'm using it—really has to do with labor and
> time; the time that you sweat and worked in that place. And that is what makes
> you feel connected to it.[8]

It is therefore because of their personal investment in the space, their freedom to decide on issues pertaining to design and usage, and because of the reflection of their selves in the space that the garden becomes *their own* backyard, *their own* piece of land. As Claudia adds: "If you feel like you put all these efforts into it, you can control more when people can come in or what kind of activities can go on in this place, even though the space doesn't belong to you, the space belongs to the city or to a private land trust"[9] then you feel you own it. It is the intensity of investment and control in the process of determination and production of space that facilitates this sense of ownership among gardeners.

The communal determination and production of the garden is the crucial process that is diverging from the experience of most contemporary urban residents. A prominent narrative of gardeners is about the transformation of "what was once a rubble-strewn, rat-infested lot into a vital green space and cultural center."[10] Therefore, having a garden "is not a matter of having a park. It is owning a park—something that we worked for and will take care of. Not some missionary that

---

7   Exceptions are Trust for Public Land gardens, in which gardeners have a collective legal ownership. This exception is elaborated on later in this chapter.

8   Interview, November 8, 2005, East Harlem, Manhattan.

9   Ibid.

10   Earth Celebrations, "Garden Preservation History: Why Are the Gardens Endangered?" http://www.earthcelebrations.com.

came in and gave us a park."[11] *Owning* the garden and investing effort and affects in it charges residents with a sense of control over the environment and provides them with an outlet for their creative needs and capacities. This interaction indeed produces a different type of ownership; it renders owners as first and foremost the producers and creators of space, and renders the "owned property" as the crystallization of the image of its producers.

Gardeners contextualize this sense of ownership within the political economy of the city: the unevenness of neighborhood development and of the resources allocated to them. Data from the 2000 census on neighborhoods abundant in community gardens suggest that gardens were not only established in the most distressed and dilapidated neighborhoods of the 1970s, but also that these neighborhoods, three decades later, still exhibit some common disadvantages. As mentioned earlier, New York City's percentage of renter-occupied housing units in 2000 was 69.8. The average percentage of renter-occupied units in Manhattan's districts in which community gardens are concentrated is 91.25, substantially higher than the city's and Manhattan's indicator of 79.9 percent. This pattern recurs in Brooklyn and the Bronx. In Brooklyn 72.9 percent of the housing units are renter-occupied, while the average of districts with a dense population of community gardens is 82.6 percent. In the Bronx such districts are characterized by an average of 93.2 percent, whereas in the borough as a whole 80.5 percent of units are renter-occupied.

As already mentioned, community gardens are disproportionally concentrated in areas of low housing ownership relative to the city as a whole but also to their residential surroundings. Since ownership of residence is directly related to health and well-being—both indirectly through factors such as tenure and control of conditions as well as directly through the meaning of home[12] for gardeners, most of whom are not owners—the garden constitutes an alternative as "their own piece of land." The process of investing in it, modifying it, and expressing their taste and culture in it instills and reinforces a sense of ownership. As Sam, a gardener from Harlem suggests:

> I think they are important also to give people a sense of ownership. Being involved with gardens really gave me a real sense of ownership. In terms of investing, it was an investment in our community. [...] But it gives people who may not be fortunate enough to own a building, it gives them a sense of ownership as well, having a piece of land. And it is really important, having a piece of land, having access to it is an important part.[13]

---

11   Johnson, K. 2000, "Green with Envy." *City Limits*, January, http://citylimits.org/content/articles/weeklyView.cfm?articlenumber=71.

12   Hiscock, R., S. Macintyre, A. Kearns, and A. Ellaway. 2003. "Residents and Residence: Factors Predicting the Health Disadvantage of Social Renters Compared to Owner-Occupiers." *Journal of Social Issues* 59(3): 527–46.

13   Interview, June 8, 2005, Harlem, Manhattan.

Furthermore, gardeners indicate that the gardens they create compensate not only for the lack of individual ownership but also for the general lack of amenities and services in the community. Gardeners refer to the lack of fresh produce in their neighborhoods that the gardens supplement; the lack of playgrounds and affordable programs for children; and the absence of parks and green open space in general. These types of "services" are even more significant in districts where residents cannot afford to travel to parks or the countryside for recreation. Ilya, a gardener in the Bronx and an activist, generalizes from his work with South Bronx gardeners and offers insights as to how this recognition may be responsible in the construction of a sense of ownership:

> […] but really the seeds of community green space is the heart of the hard-working people who were there originally when the place was bulldozed, burnt down, everyone left and it was the people themselves saying "We need to make it a better place for our community and *build ourselves for ourselves by ourselves.*" No government nobody was helping at all; actually they were taking away the normal things the community had, like trash pickup etc. etc.[14]

Thus, as the last two gardeners suggest, sense of ownership is not only being formed through the labor that is constantly invested in the space and the control that is exercised in it, but is also being fortified by the social context in which these occur. Residents of a neglected space, a no-man's-land, were forced into independent, unsupported struggles for survival in which the garden was both the salvation and victory of the community as a whole. This reclamation of ownership over their life that was afforded through producing the garden entails not only the determination of space but also self-determination—"build ourselves for ourselves by ourselves"—and makes this common space perceived as their own.

A strong sense of ownership, one that cultivates affects and responsibilities towards the environment, cannot be induced "top-down," but depends on a process of production: it is necessary to be personally invested in the space. Indeed, for the most part gardeners are free to design a space that will satisfy their aesthetic, personal, and communal needs, and manage their product with almost complete control (except for the very basic guidelines provided by Green Thumb). This understanding can contribute not only to how we perceive and work with community gardens but also with other community institutions that depend on high levels of voluntary participation. This understanding is now more widely accepted among organizations that oversee the gardens, such as the city's Parks and Recreation Department, and grants gardeners some protection from top-down planning, design, and operational impositions, which in turn helps maintain the gardens as grassroots-produced spaces and as spaces of free expression. However, this "noninterventionist" policy may be driven by a more rational calculation:

---

14   Interview, May 31, 2005, South Bronx.

there is a very small budget allocated to Green Thumb to support the gardens, while the city enjoys voluntary maintenance work of urban green open spaces.[15]

The words of the Assistant Parks Commissioner, Jack Linn, responding to gardeners' questions in regard to further funding for gardens, capture the duality that is in play in the governance of community gardens:

> We don't want to spend a lot of money recreating something that you created, and building something that Parks [Department] creates—that is not a garden. Your sense of ownership is lost and we end up having to maintain it as we maintain a playground. This is the opposite of what you do, which is community building.[16]

At this historical juncture in the city's political affairs there is willingness to endow residents with more power and greater control over the public land of community gardens. At the same time, the municipality neglects to provide the gardens with services that are provided to public parks, such as water connection, electricity, and toilets.

Two Models of Participation

This characterization of community gardens in New York City, however, no longer applies to the 126 gardens that were purchased by the two nonprofit organizations— the Trust for Public Land and the New York Restoration Project—to prevent their sell-off for development. By purchasing the gardens, the nongovernmental organizations (NGOs) inaugurated two new models of urban community space with new meanings of community gardens. The discussion that follows compares these two new models of community gardens, with a focus on the generation of a sense of ownership, control, and participation among residents. These examples suggest, among other things, that there is no one all-encompassing approach among the various overseeing organizations regarding the meaning of community gardens and the relations between residents and this specific urban environment.

NGO management of public parks (preserving them as "a land bank" for future generations) represents a private mode of production of nature, "an accumulation strategy."[17] As "'preservation' is most commonly accomplished

---

15   Rosol, M., 2012. "Community Gardens as Neoliberal Strategy? Green Space Production in Berlin." *Antipode* 44(1): 239–57. Rosol examines community gardens in Berlin and identifies this municipal approach as outsourcing of public services (that is, maintenance of green open space) to voluntary forces.

16   Jack Linn, Assistant Parks Commissioner, April 22, 2006 at the Second Annual Gardeners' Forum.

17   Katz, C. 1998. "Whose Nature, Whose Culture? Private Production of Space and the 'Preservation' of Nature." In B. Braun and N Castree (eds). *Remaking Reality: Nature*

by a physical and textual exclusion" of (local) users it redefines the value and usage of space and challenges its potential to evolve as autonomous community space.[18] This autonomy is "hinge[d] on the ability of those within localities to *appropriate control* over the structures and institutions that are the focal points of those [mundane] *experiences and practices.*"[19] Autonomous community space, then, entails a high level of residents' participation in managing, influencing, and (co)determining their environment. Autonomous community space helps accentuate the multiple potential meanings and forms embedded in property, while land preservation through NGO intervention could be seen as part of the prevailing "property regime."[20]

Model 1: Land to the People—the Trust for Public Land

One model of nongovernmental management of community gardens was developed by the Trust for Public Land (TPL), which was established in 1972 as "a national nonprofit that conserves land for people to enjoy as parks, gardens, historic sites, rural lands, and other natural places, [...] ensuring livable communities for generations to come."[21] As a national organization, TPL purchases land and sells it to government agencies. In its urban land program, however, the lots are managed by local groups as community gardens or mini-parks.[22] Between 1978 and 1999 TPL sporadically purchased 10 successful community gardens across New York City, each becoming a "single-site land trust." Gardeners own these gardens and manage them independently. Out of the 10 gardens, five are still active and thriving. In 1999 TPL purchased 64 gardens that were slated for auction, and since then it has purchased or received as donations five more gardens. TPL has since worked to transfer the legal rights of its 69 gardens to the gardeners. In 2011 TPL transferred the legal rights of its Bronx gardens and Manhattan gardens to the Bronx Land Trust and Manhattan Land Trust respectively. The transference of the legal rights to its Brooklyn-Queens gardens was likely to be completed in 2012.

TPL intentionally purchased endangered gardens that exhibited a somewhat developed and participatory organizational structure. The executive director of TPL in New York reviewed TPL's logic at the time:

---

at the Millennium. New York: Routledge, p. 47.

   18   Ibid., p. 54.

   19   DeFilippis J. 2004. *Unmaking Goliath: Community Control in the Face of Global Capital.* New York: Routledge, p. 10 (emphasis mine).

   20   Blomley, N. 2004. *Unsettling the City: Urban Land and the Politics of Property.* New York: Routledge.

   21   http://www.tpl.org/about (accessed December 5, 2011).

   22   Brewer, R. 2003. *Conservancy: The Land Trust Movement in America.* Lebanon, NH: Dartmouth College Press.

The gardens we chose were the more active of the gardens that were for auction. It was not the TPL's philosophy to buy land and be sort of a *land bank* [for] land that *might* at some point be run by these communities. We wanted to acquire gardens that had some sense of organization in operating them. So we chose ones that were more or less working [with] leaders [who] were quite active and we knew, it was our vision anyway that, as much as feasible, the decisions about community gardens' future should be made by the garden volunteers. That is the way community gardens emerged in New York City. Community gardens didn't start based on an agency starting them off. They grew up from the grassroots and so we did have a philosophy that they should continue to be grassroots-run.[23]

TPL's model echoes the original characteristics that produced community gardens and their original meaning—grassroots, decentralized, co-determined, and autonomous space—and wishes to reinforce these by transferring ownership to the gardeners. TPL therefore invests efforts in networking gardeners within and amongst gardens and in decentralizing powers in order to establish a unique model of legal collective ownership for community gardens.

TPL works with the gardeners to facilitate as participatory a model of operation as possible, in which gardeners are the dominant decision makers responsible for their gardens. In order to do so, TPL facilitated meetings with its gardeners citywide and between its gardeners within each city borough. These efforts resulted in the assembly of a board of directors to represent all of TPL's gardens. The first mission of the board of directors was to configure the form of their own management. After much deliberation the board decided to establish three separate land trusts to manage TPL's gardens in each city borough (Bronx, Manhattan, and Brooklyn-Queens) according to their special needs and interests. Once the three trusts were established and functioning, TPL could transfer the property entitlements to each of them. This step endows gardeners with real (that is, legal) ownership over the spaces. Each of the land trusts will be responsible for the properties, manage the budget of the gardens in its respective borough, and be self-sufficient through fundraising. According to this model, participation (through a high level of community engagement in the production of space) is the quintessential component of community gardens. It is far superior in importance to merely preserving urban land as open space.

TPL's gardeners are entitled to collective ownership of the space as long as they maintain a collective and act as an inclusive community resource. By transferring the gardens to land trusts, TPL ensures that the property is taken out of market competition, albeit not permanently. Gardens' sustainability depends on the community interest in having a garden. As long as residents maintain and use the land as a garden, as long as it serves the common wealth, its protection from market forces is guaranteed. Indeed, TPL's emphasis on community participation implies that land can possibly be reappropriated and redesignated in cases of neglect and

---

23   Interview, October 17, 2006.

lack of community participation. In this model, community gardens are perceived first and foremost as a *community* resource. In order to help the gardens function as a resource for the common wealth, TPL is highly invested in community building and organizing. A representative of TPL in New York explains:

> TPL decided [...] to focus at least as much on the organization as on land tenure because permanent land tenure is meaningless if the organization can't sustain itself [...]. There are colleagues of mine to whom an empty garden and a vacant lot are things of great beauty and value because it is open space for the future. And I'm much more focused if the community isn't invested it may never be invested so you['ve] got to keep it invested and you['ve] got to do what you can to perpetuate the organization, to encourage and help it be public or else whatever land tenure you have isn't worth much.[24]

For community gardens to serve a wide variety of public purposes and needs "it is crucial that there will be an organization that can run [the gardens] for public benefit [...]. Community gardeners will play a significant role in governing the organizations and the gardens will become increasingly important for their neighborhoods as a result of being as public as possible."[25] According to TPL then, if the space is underutilized and its resources are no longer needed and appreciated by the community, it loses its real purpose and should no longer be protected.

The new model that is being developed in TPL's gardens creates a new meaning of community gardens. This model emphasizes community participation above and beyond protection of the land itself. According to the model, *participation* is the essence of community gardens, which cannot be sustained without it. It remains to be seen how this new realm will affect the perceptions and behaviors of gardeners. While solving many problems, this model also creates new challenges pertaining to ownership for the gardeners. As a representative of the Council on the Environment of New York City (CENYC) analyzes:

> The gardens and the land trust groups are now realizing that owning the property is a really big responsibility. So I think over the years it was always the thought that if the city will own the property, let's not have the gardeners have to go through that, have to deal with all the ramifications of owning property. When there was no other option, you know when the gardens were auctioned off, then it was like, well, now we have to change our thinking a little bit, and yes, OK, let's try to get as many gardens to land trusts if the option is they are going to be lost. So now the whole idea of land trust is another way towards sustainability.[26]

---

24 Ibid.
25 Ibid.
26 Interview, December 5, 2005.

Gardeners from various TPL gardens report on the massive amount of work and challenges they had to address in the process of reinventing this modality of community gardens. They had numerous meetings and vigorous discussions and deliberations to navigate between different visions in order to develop their management format. For some it was too much. After going through the struggle of protecting the gardens, they wanted to be able simply to garden again. Those who took part in the process spent a lot of time in meetings to establish boards of directors for each land trust, and learned what management of the land entails and about the various needs and difficulties of garden groups. This process demanded that gardeners develop a whole new set of skills and knowledge, far beyond what was required for the management of their own garden and for activism in neighborhood organizations and coalitions. Billy, a land trust gardener in his fifties from Brooklyn, explains:

> I was very involved in setting up the Brooklyn version of Land Trust, the Brooklyn-Queens Land Trust ... people like myself got stuck in going to Land Trust meetings that are way too detailed ... ho my god, why are we reinventing the wheel—the wheel is already invented. This is the way they are going to do it? They are going to torture us ... Now they have this board of directors making decisions on everything from membership to funding your garden party. They are so skewed in what the real board of directors should do. It is not working well. It is not working. I put up with that as long as I could.[27]

Thus, legal ownership requires gardeners to be involved in the administrative management of the land trusts *and* their property. In some cases, this kind of involvement impedes rather than encourages ongoing participation in the production of community space. The spontaneous collective action of residents over space is repositioned within a rigid structure of management and responsibilities that diminishes the grassroots essence of community gardens and its dynamic and organically developing production.

The process of establishing independent land trusts encountered further difficulties as a result of TPL's desire to construct the board of directors in the same spirit as its vision for the community gardens, most notably by making the gardeners the owners of their gardens and endowing them with a position of power that they did not usually occupy. TPL struggles to occupy the board of directors of the land trusts with a majority of community gardeners, contrary to the commonly held perception that a board should be populated by powerful, well-connected, and experienced people. A representative of the organization describes his deliberations:

> A major difficulty is grassroots leadership, mostly from low-moderate income neighborhoods, dealing with a particular kind of responsibility, responsibility

---

27   Interview, July 29, 2006, Crown Heights, Brooklyn.

for owning all this land that has all this public use; it is a weighty burden for grassroots volunteers. And that is one of the reasons that people usually end up having land trust organizations where, at best, you have a small minority of board members being community gardeners because it is better to get people from planning departments, real estate companies, and foundations to oversee it. They are used to sharing this fiduciary responsibility, making this kind of decision and working together on boards, as opposed to community gardeners that mostly just want to garden.[28]

TPL is highly committed not only to transferring ownership to the gardeners but also to transforming the prevailing modality of property and power entitlements. Therefore, as part of its mission TPL wishes to place the gardeners—usually marginalized and disadvantageous residents—on a better playing field within the social constellation of power in the city and make them part of the "haves" rather than the "have-nots." But TPL's model of legal collective ownership puts many new responsibilities on residents. TPL "capitalizes" on the sense of ownership and sense of control that residents developed towards their gardens before they had any legal entitlement to the land, and wishes to produce a modality of public space based on institutionalizing gardeners' sense of ownership as full legal ownership with their full participation and control.

Model 2: High-Quality Open Space—The New York Restoration Project

A very different management model of public space is employed by the New York Restoration Project (NYRP), a nonprofit organization that was founded in 1995 by actress Bette Midler. In 1999, NYRP purchased 51 community gardens that were to be auctioned for development. Since then it has purchased six more gardens. Unlike the Trust for Public Land, NYRP was not looking for the most active gardens (those with an organized and democratic leadership). Some of the gardens it purchased had problems of dictatorial leadership (with one person deciding everything) or were neglected in the face of impending evacuation.

After purchasing the gardens, NYRP decided to redesign them according to a rationale resting on another kind of attachment—that between a donor and their investment rather than between the gardeners and their affectional investment. NYRP linked each of their gardens to a foundation or a sponsor, hoping that capital investment in a specific space rather than general monetary donation to community gardens would encourage contributions. In this way, donors were able to follow up and see exactly how their money was being used.[29] In some cases, the

---

28   Interview, October 17, 2006, TPL office, Manhattan.
29   Interview with NYRP's representative, November 30, 2004.

gardens were renamed after the sponsor[30] and the sponsor was granted the power to choose the layout or theme for the redesign of the garden.[31]

Professional landscape designers were recruited, sometimes through competitive tender, and money was allocated to transform existing gardens into well-designed, highly invested green areas. According to the organization's website, "NYRP seeks out innovative designers who are on the cutting edge of open-space development and green design. These partners bring creative, beautiful and sustainable design solutions to our park and community garden restorations."[32] Much-needed amenities were installed in these gardens, such as water tanks, electricity, and toilets, which generally would not be found in other gardens.

NYRP's representative, elaborating on the rationale of the organization, suggests that in order to prevent the gardens from going through their usual "organic cycle," which includes periods of deterioration, when leading figures in the garden leave or when the community loses interest in the garden, NYRP seeks to be fully accountable for the gardens and to maintain them as beautiful green public spaces.[33]

The whole redesign process, from the selection of the designer to design and construction, was executed while excluding the gardeners themselves. One of the salient results of this process is that the people who previously tended to these gardens felt detached from them. A female gardener in her thirties from NYRP's garden in Harlem recalls that NYRP's approach made her community feel as if:

> [NYRP was saying] "this is our property and you are guests on it." And it didn't feel good for me personally, it was really difficult. We lost garden members because of it, 'cause they say, "we can't garden here because we don't feel welcome in this place. We feel like at any time NYRP will change our design or want to do something different and we can't be here."[34]

In some cases the new design did not address the needs of gardeners, such as in gardens that used to focus on food production and were redesigned with open lawns, seating areas, and small flowerbeds. One gardener in the East Village lamented the loss of a particular plant—a black rose bush that had taken years to cultivate—which had been destroyed by NYRP's staff in the process of "clearing up" the space for redesign. In another NYRP garden in the Lower East Side, Suffolk Street

---

30   For example, Target Community Garden and the Heckscher Foundation Children's Garden in Brooklyn; and the Riley-Levin Children's Garden and Rodale Pleasant Park Community Garden in Manhattan.

31   For example, the Family Garden, sponsored by Tiffany & Co., and the Ann Richards Garden in Manhattan.

32   http://02b4300.netsolhost.com/greening/index.php?sub=2&print=yes (accessed March 1, 2011).

33   Interview, November 30, 2004.

34   Interview, November 8, 2005, East Village, Manhattan.

Community Garden, the casita—the focal structure in Latin American gardens— was "barricaded" by three Victorian-style stone benches named by Bette Midler after her friends: Sarah Jessica Parker, Mathew Broderick, and another celebrity (see Figure 3.1).

Figure 3.1   NYRP's Suffolk Street Community Garden

The top-down exclusionary process of recreating NYRP's gardens detaches the gardeners from the space they used to perceive as their own. After the redesign, the gardeners experienced the areas as a public park: a joyful experience of open space, greenery, and nature is still possible, but experienced from a distant—as an outsider, a visitor, a guest rather than as a member-owner. It is not a space that reflects its users' identities, cultures, and sense of aesthetics; it was not produced collectively by residents according to their style, desires, and needs.

Another NYRP garden that was completely redesigned by an outside contractor without the involvement of the neighborhood is now perceived by the residents as a display of greenery or, as one of its previous gardeners put it, a "flower museum"—again an indication of the gardeners' detachment from it. In a visit to this garden in Harlem I encountered a group of elderly African American women sitting on folding chairs on the sidewalk in front of a closed-gate garden. This "Family Garden" was sponsored by the Tiffany & Co. Foundation and themed on Thomas Jefferson. Tiffany's designer, John Loring, introduced elements such as:

a wrought-iron entry gate that echoes the façade of Jefferson's beloved Virginia home, Monticello, and garden furnishings that feature the Founding Father's silhouette. Other Jeffersonian images and Colonial-era features are found in cut-metal figures affixed to the garden's planting beds, a bluestone patio, a traditional arbor and brick planters with wide limestone caps.[35]

This description of the design neglects to mention that the heavy iron chairs are fixed to the ground and that the brick garden's beds feature flower displays in a garden that was formerly used for growing vegetables. While this garden is not used by the residents, a neighboring one has a long waiting list for vegetable plots. When asked why they were not inside the garden, the old ladies silently nodded and smiled sadly. The sight of former gardeners sitting outside the locked gate of the garden was testament to the breach in their sense of ownership—they were not completely detached but already set apart from it. A gardener from a neighboring plot describes the difficulties of the Family Garden:

> They put those chairs in there but the chairs are fixed. It is not in any way accommodating or welcoming for the community to use this space. And really all they do is sit in front of the garden. I've never seen people sitting in the garden. I have seen people in the garden looking at the flowers, it is kind of like going to a museum, "how cool, look at those flowers." But what's going to happen in the end is that NYRP is going to have to pay their staff to go in and maintain the space because nobody feels necessarily it is theirs to go in and work.[36]

The operation of some other NYRP gardens is more successful, although they also went through many difficulties in the process of transferring to NYRP. Rene, a gardener from one of the more successful NYRP gardens—Pleasant Park Community Garden in Harlem—recalls her experience of disappointment, frustration, and struggle during its transformation into a NYRP garden:

> The redesign process of Pleasant Park ... for me, personally, was very difficult and really made me realize how attached I felt to this piece of land even though it wasn't mine ... I was disappointed for not being included more in the design process. And I feel like we as a garden group have spent three years discussing and arguing amongst ourselves what we want to build in the garden before this thing happened. And we have come out with a design that took a lot of work for us. And that design was not even looked at, and they hired a landscape designer from outside of the city to come up and do the design.[37]

---

35 http://www.nyrp.org/Parks_and_Gardens/Community_Gardens/Manhattan/Family_Garden_Sponsored_by_Tiffany_Co (accessed March 1, 2011).

36 Interview, November 8, 2005, East Harlem, Manhattan.

37 Interview, May 6, 2006, Harlem, Manhattan.

Despite these grievances, on a practical level NYRP and the garden group reached a compromise regarding the design; and the gardeners' stubborn efforts to become involved in the design process effected the "rules of engagement" between this group and the organization towards possible dialogue in cases of concerns or suggestions.

NYRP manages its 57 gardens as reservoirs of urban green land that the organization is responsible for securing and maintaining. However, this model of management of public space undermines the quality of residents' interaction with the space of the gardens. The space that gardeners received after the redesign of their gardens felt too distant for most of them to reengage with. Without their personal investment in the space, it no longer reflected their ideas of aesthetics, self-image, and culture and did not respond to their needs. Ironically, the new space was a sad reminder of what was destroyed—the old garden. By eliminating the personal investment, care, and involvement of gardeners, NYRP's model deprived gardeners of a sense of ownership and control over the space. Claudia from NYRP's garden reasserts:

> We feel like at any time NYRP will change our design or want to do something different and we can't be here. Personally, I think for them it was not a good move because if you want your gardens to look nice—they consider the gardens their property and they want them to look nice so that they can raise money to support the organization—the best way for those gardens to look nice is for the gardeners who garden them to be happy. And how do you do that except by listening to gardeners? Yes, it might have made the entire process slower. The redesign of our garden could have taken three times as long because there will be meetings and listening and blah, blah, blah and yes, it would have been annoying and not very efficient, but I think the gardeners would have felt more invested in the space in the long run perspective.[38]

NYRP's model of public space management established a new conceptualization of community gardens, stemming from a rationale of centralization, efficiency, and land preservation. As a result, both gardeners and the organization perceived these "community gardens" as the *property* of NYRP, which employs professional staff to design and in many cases (owing to lack of community participation) maintain the property. Since *community* is not a leading rationale, NYRP is also making no effort to forge connections between garden groups and to create a collective of NYRP gardeners. One consequence of such an approach is that NYRP gardeners usually refrain from taking part in the citywide gardeners' collective. The gardens then become a version of private-public parks that are maintained by professional staff.

NYRP conceptualization of community gardens established new modalities of interaction between local residents and the gardens; between the formers and the citywide collective of gardeners; and between the overseeing organizations, the space, and the community. A gardener from NYRP clarifies:

---

38   Interview, November 8, 2005, East Village, Manhattan.

NYRP [...] as far as I seen, made no gesture to talk about gardeners having actual ownership over the property. I don't think they have any intention of doing that. And also they made no effort to connect their gardeners to each other. Where the TPL did all this work to get gardeners together from different gardens, [...] NYRP owns gardens that are three blocks from my garden and [...] me knowing them has to do with me going to introduce myself because there is no effort on NYRP part to get us together.[39]

The NYRP model for garden preservation dramatically changed the way that gardeners perceived the garden and interacted with it. The garden is no longer experienced as a communal shared space that evolves over time (like the vegetation in it), like a collectively determined and objectified space: "It is not like an organic thing that was built over the years where people were piecing together little bits from here and there. [It] definitely doesn't feel like anybody's space that lives on that block."[40]

NYRP's approach has changed the meaning of gardens in a fundamental way. Since the NYRP model focuses on land preservation rather than community participation, in many cases the community is excluded from the process of production of space and from the space itself. In addition, this model of protecting public space is highly centralized, self-determined by the organization, and therefore highly dependent on its resources. The NYRP model, then, diminishes the grassroots political-organizational infrastructure of community gardens, and with it dissolves their opportunity to develop as autonomous community spaces.

To sum up the comparison, throughout their existence, until the New York Gardens' Settlement of 2002, the concept of preservation as legal protection was not applied to community gardens. Since they had no legal protection, gardeners had to renew their annual leases knowing that the gardens might be taken away at any moment. Despite this general instability, gardeners exercised high levels of participation in and control over space, developed a strong attachment to their gardens, and experienced psychological ownership—all of which enhanced their struggle against the intention to auction off many of the gardens. A central achievement of the struggle over the gardens in the face of hungry market forces was in situating the issue of gardens' sustainability at the heart of public and political discourse. NGO management of public space was introduced as part of the solution.

TPL and NYRP invented new meanings of community gardens, presenting two highly differentiated models of management of the gardens with different interpretations of the role of residents—the community—in participating in, managing, and co-determining them. The new concept of community gardens under NYRP, which evolved out of an effort to save the gardens from the private market, paradoxically replicates the modern private urban space that excludes residents from the process of its production. Though not intended for purposes

---

39   Ibid.
40   Ibid.

of capital accumulation, the NYRP model is based on instrumental rationality and centralization of power, components that characterize the workings of the private market. Garden groups are atomized and to a large extent excluded from participation in the production of space. As a result, the significance of the garden as a space for participation, self-expression, and cultural celebration—and as the bedrock underlying residents' sense of familiarity, belonging, and ownership—is significantly undermined. It is a type of public space that offers recreational activities in a very structured and safe format through suppressing expressions of multiplicities of design, usage, and users.[41] Such a model contributes high-quality and well-maintained open spaces to urban neighborhoods, but fails to build the social infrastructure that would allow the making of such space into an inviting, inclusionary, and relatively autonomous community space. NYRP is concerned with preserving community gardens, but its focus is not on the community as it is, in fact, preserving the gardens as open land reserves for the future—an accumulation strategy that necessarily excludes local users.[42]

By contrast, TPL focuses on community participation through a well-established community organization and on the creation of autonomous community open space, thereby facilitating residents' sense of ownership and legalizing it. But in so doing, TPL transfers also the responsibility for managing a public space to the residents. Legal ownership adds many more bureaucratic and organizational challenges to the daily operation of the gardens. Therefore, beyond the many voluntary hours resident-gardeners spend maintaining their gardens, in membership meetings and work days, and in building strong protective networks of collaboration at neighborhood and citywide levels, they now have to take on extra duties that were formerly the responsibility of the municipality.

The model for sustainable community gardens that is being developed by TPL may be subject to two conflicting interpretations. On the one hand, it assures a true collective ownership of space in which gardeners are the owners and decision makers and therefore do not have to depend on outside organizations or be affected by changes in local politics. On the other hand (similar to the NYRP model) TPL presents a conceptual challenge to the idea of autonomous community space. In the TPL model the gardeners, in line with the tenets of a capitalist society, own the garden. Such a model reaffirms the importance of ownership *per se* as the only means of protecting public space from market forces and at the same time diminishes the responsibility of the "public"—city agencies and local politics.

Community gardens depend on mechanisms of participation, control, and sense of ownership in order to develop as autonomous community spaces. The NYRP model suppresses many of these mechanisms, and in some cases eliminates them altogether. The TPL model intensifies these mechanisms, but in doing so also changes the realm in which the community (spontaneously and effectively)

---

41   Mitchell, D. 1995. "The End of Public Space? People's Park, Definitions of Public and Democracy." *Annals of the Association of American Geographers* 85: 108–33.

42   À la Katz 1998.

participates in public space. The dissolving capacity of the state to protect public assets from market pressures transfers much of the responsibility to the realm of civil society. The two models can, in fact, be understood as the two faces of civil society. One conceptualization of civil society, associated with conservative ideology, sees its function primarily in correcting and supplementing the ill-treatment and incapacities of the state while maintaining the status quo. The other conceptualization, associated with progressive and radical politics, sees civil society as an autonomous social platform that mitigates the power of the state and the market.[43] The analysis presented here suggests that the NYRP model can be associated with the former, while the TPL model is better aligned with the latter. However, the solutions that civil society can offer are problematic at best and in both cases change the meaning of community gardens and public space. The two solutions that TPL and NYRP provide to the increasing scarcity of open public space in inner cities are based on civil-society organizations, backed by private money, with the goal of reversing a public decision to privatize the space. But for many years the municipality was relatively successful in overseeing these community spaces while enabling residents' participation without loading technocratic management duties on them. A third conceptualization of civil society, which sees its role as protecting democracy and making sure that public institutions work for the benefit of their community, may better serve the need of residents for autonomous public space.

Found Space: Discovering Opportunities

> Men can see nothing around them that is not their own image; everything speaks to them of themselves. Their very landscape is alive.[44]

Henri Lefebvre suggests that the modern capitalist landscape lends "the city as a product" rather than "the city as work, as oeuvre."[45] Following Marx, Lefebvre upholds the significance of space in shaping society, in being both a product and a reproduction mechanism of society. But in the modern urban landscape, as people understand themselves through their reflections in it, important components of their being are alienated.

The space of the garden offers a different reflection: it serves as a connective tissue between past and present; among segments of identities of individuals; and between one's sense of self and sense of place. These meanings of space are

---

43   Cox, R. 1999. "Civil Society at the Turn of the Millennium: Prospects for an Alternative World Order." *Review of International Studies* 25(1): 3–28.

44   Karl Marx in Burgin, V. 1996. *In/different Spaces: Place and Memory in Visual Culture*. Berkeley: University of California Press, p. 2.

45   Isin, E. (ed.). 2000. *Democracy, Citizenship and the Global City: Governance and Change in the Global Era*. London: Routledge, p. 14.

enhanced by the level of control that people are able to exercise and by their sense of ownership over the environment. Moreover, gardeners appreciate and cherish these qualities of the garden precisely because they are not easily obtained in the urban environment. For many individuals these meanings and experiences of space are a leverage and springboard to other ways of being.

Gardeners' environmental autobiographies provide bountiful insights into how new opportunities are found and pave the way to what Heidegger termed another way of "being in the world." It has already been suggested that the garden is well integrated into the life stories of gardeners and is a major component of their environmental autobiography (Chapter 2). This is also manifested in the narratives of gardeners regarding how they became gardeners. They frequently introduce the narrative of their becoming a gardener when portraying the importance of the garden to them: that is, the manner in which they discovered that the garden was significant to them. It may suggest that becoming a gardener brought dramatic changes to their life, significant enough to remember and integrate into their life story.

Becoming a Gardener: Reactive and Proactive Narratives

Two overarching narratives on the process of becoming a gardener, differing in their levels of agency, can be distinguished: the reactive and the proactive. The reactive narrative tells the story of being "recruited" or pooled into membership in a garden. This could be attributed to the influence of a leading figure in the community, such as the "Christian brother" who invited a friend to join in—as the story of Mr Thomas, a gardener from East New York (originally from a housing project in upper Manhattan) reveals:

> I came along because one of my Christian brothers who I've known from 1998. He has a garden about three blocks down and I ran into him one day and ever since then I was part of that garden. [...] And there is a whole world within the greening community which I had no idea about when I first met my Christian brother with his garden. I was thinking oh we got all the space I can get cucumber [...] no, it is much bigger than that. And slowly I began to learn a little more because I'm not a gardener at heart. I don't know anything about this. I'm concrete. I'm the city guy. But I do love fresh fruits and vegetables and I love the earth I love the greenery I love the quiet [...] like what we are doing now I can do it forever. And I began to learn from other gardeners [...].[46]

The "recruiter" could be an activist who inspires new people to get involved—such as the legendary Françoise Cachelin, a leading figure in the gardens' movement until her death in 2003. She lives on in the stories of many gardeners as the person who initiated their journey in community gardens. Sofia, a gardener from the East

---

46   Interview, May 6, 2006, East New York, Brooklyn.

Village and an active member in the New York City Community Garden Coalition (NYCCGC), exemplifies:

> I moved into the neighborhood in 1984, so right when the gardens were flourishing. Then Françoise and I became friends and because she was so overwhelmed, it was during [the time of Mayor] Giuliani, when a lot of the gardens were being lost to developers, and some of them were so beautiful it was literally a crime to see them go. [...] So she basically asked me to come, they planned a very big event called "Standing Our Ground" and she asked me if I want to help out. That was 1997. So I told her I will help out whenever I could. [...] That is how she got me involved. And she wouldn't take no for an answer, she would call and said "If you don't come the gardens will be gone" and she had that kind of force she could convince anybody.[47]

Another "recruiter" could be an organization that through its activities with the gardens new people discover them and join in. One such organization is Little Sisters of the Assumption, which brought a group of Mexican female immigrants to a local garden in Harlem as a food relief and skills and leadership development program. According to the coordinator of the group, the women initially followed the program because they needed the substantive support it promised, but they were somewhat reluctant. The coordinator recalls:

> Within my first year I realized these women don't want to take control over the project. They just want to go and do work and have somebody else really in charge of everything and actually hold all the power. They actually set me down at one point and said: "We don't want to have these discussions about consensus," "We don't want to have you ask us what to do, about when we are planting," or "What is the public access to this space going to be?" or all the questions that come up in a community garden. "We would rather you decide these things and tell us what it would be." [...] I think it is hard for somebody in that position who not only doesn't have a lot of education but now find himself an immigrant in a foreign country. They are not someone who will necessarily step up and say I am going to take control of this project because they are in a position that does not hold a lot of power to begin with.[48]

However, this group, like other groups, went through some processes that sparked a different approach. The coordinator reports:

> But I found that some of these women are actually natural leaders in a different way than I expected. Not in a way of standing in front of the group and talking about their opinions but more in a way of networking slowly with the women

---

47 Interview, October 7, 2005, East Village, Manhattan.
48 Interview, November 8, 2005, East Harlem, Manhattan.

on a one-to-one basis outside of our meetings. They now form a system where they have bylaws for how our group functions and how people can participate in the group, and they have a president and a vice president and a secretary of the group. And the woman that is now the president of the group is one of those that can't read or write but she is actually an excellent leader and she is taking on a lot of the responsibilities in the garden in a way I didn't really expect. [...] And [they] started a small farmers market. They talked about it last winter and I thought they will start in the spring but [...] it took them a long time; it took them until July to get organized. But they finally did and like August, September, October they had six or seven farmers markets.[49]

The collective production of space enabled them to manifest and develop capacities and change their reaction to the environment from avoidance and anxiety into competence and agency.

The second narrative, the proactive one, presents the process of becoming a gardener as a conscious ideological step, construed as part of the general ideology of the person and as the natural progression of the person's life and activities. Within the active narrative we can distinguish between two levels of awareness. The first level of awareness in the proactive narrative is a local, problem-specific consciousness that characterizes mostly the first generation of gardeners. These are residents who came together in order to deal with rubble-strewn abandoned lots adjacent to their homes that became breeding grounds for drug dealers and prostitutes. These socially concerned and active residents took matters into their own hands and initiated the cleanups of the lots, transforming them into green and safe oases for themselves and their communities. This proactive narrative presents a struggle for survival in which the underprivileged were able to transform their conditions in the city through the opportunities that they found and established in the space of the garden. The words of one gardener poetically describe this force of survival:

> For me it all started, the revolution started with the first little old lady from Puerto Rico that planted those seeds of sunflowers in the [space that is now a] garden and saw that there was potential even in the most toxic cities, where the land fall in on itself, and burnt down with lead and this and that.[50]

The second level of the proactive narrative—urban/social consciousness— presents community gardens in broader social and political terms. According to this narrative the gardens are significant beyond providing a solution to a specific problem of a specific urban block. Rather, involvement with community gardens is contextualized within the neighborhood, the city, or society as a whole. The proactive narrative focuses mainly on issues of open space and urban development,

---

49   Ibid.
50   Interview, May 31, 2005, South Bronx.

but in many cases it stretches to social problems at large such as poverty, health, and environmental problems, and uneven urban development (Chapter 5 elaborates further on these subjects). The gardens are perceived as sites through which it is possible to address these problems. Ilya, a founder and activist of the More Gardens! Coalition,[51] describes what propelled him into action. In his story, art and the promotion of a certain social agenda found the right platform in community gardens and evolved into activism:

> When I got to NYC as a photographer and an activist, by that time working with schools doing projects, I decided that art is not just for pretty white spaces where few people get to see it but it was part of activism, community and enhancing life. [...] And then from that coming to community space Plaza de Sol, which then became endangered. I learned about how community gardens work. And there was the Earth Celebration which I really like and worked with for a while and that also gave me more leeway into how community gardens work, but that was it. I was doing performance, puppetry in these community gardens about preserving them.[52]

The story of another gardener-activist manifests another facet of the urban/social consciousness. For Sam, the gardens were part of community building, a mission that he also struggled for by other means. Community gardens in the city have presented themselves to him as an opportunity to execute his vision of equitable and just land use and community power in yet another way:

> I got involved because it was part of something I had worked for with "Brooklyn Core" [Congress of Racial Equality] on housing discrimination in Brooklyn and also with the National Black United Front. We were trying to reclaim some buildings in Brooklyn and to work on them and turn them into livable affordable housing. But even before, I worked for the Black Panthers party in the 70s. And whenever I come across a project that looks like it was community oriented working in grassroots I tried to get involved.[53]

---

51 More Gardens! Coalition is a nonprofit organization established in 1998 to help endangered gardens and to fight for more gardens in New York City. The coalition does community-based work to organize and educate gardeners and the community; to file lawsuits; develop alternative plans for neighborhoods; and to raise public awareness and funds to help gardens. Initially, it was based in the Lower East Side where, among other activities, the coalition organized a nine-month encampment in the Esperanza community garden. Later, the coalition moved its base to the South Bronx and began working with numerous endangered gardens to propose alternative development plans for the areas. Later on the coalition moved its base to East Harlem, working with threatened gardens there.

52 Interview, May 31, 2005, South Bronx.

53 Interview, June 8, 2005, Harlem, Manhattan.

Another example of the proactive narrative comes from a representative of CENYC. In this case working with the gardens is connected to environmental awareness and urban development:

> I have a degree in social ecology [...] and that is kind of the study of ecology and organic gardening and alternative energy and technical things like that, but with the understanding that the social part is really a key too. You can have the answers to all the technological questions in the world, but if you can't get people to agree or come up and work together on solving a problem or just living together then none of that stuff is useful at all. So I think when I began to look for work in the field I felt that the city is a very important place for dealing with these issues and I also felt that gardening and the environment in the city is also a key, real key part of making the city livable.[54]

Ample opportunities are found in the gardens, whether gardeners approach them proactively or reactively. Looking at individuals from the inside out, from the personal to the social, the most immediate experience of the space by gardeners includes finding those personal opportunities to reconstruct their past landscapes; to reenact past practices; to express their sense of aesthetics and diversify their aesthetic experiences; to cultivate a space for celebration and recuperation; to belong; to better integrate their living environment to their life stories and selves; and to experience ownership and exercise control over this environment.

The reactive and proactive narratives differ from each other mainly in the spatial understanding that was exercised in those initial moments of joining the garden. It may be that gardeners who narrate this moment reactively were less likely to see many of the affordances and opportunities that the space envelops and were more influenced by other mechanisms to join in. The proactive narrative, on the other hand, indicates awareness of at least some of the potentials of space to bring about change to the lives of gardeners and the community at large. It is interesting and important to mention here that the type of narrative that is expressed in regard to the "initial moment" does not in any way determine the future understanding of opportunities. The variation of narratives, however, helps to expose how spatial understanding and political consciousness develop over time and through specific kinds of social and spatial practices: that is, the development of spatial gaze and spatial practices entailed by the movement from a reactive level to a proactive level and between proactive levels (from the problem-specific to broader urban consciousness). The result is a more complex, well-rounded, and politicized consciousness of residents. It is a process in which alternatives to the hegemonic discourse can arise.

With these found opportunities we can better understand the connection between practices of resilience, practices of reworking, and practices of resistance (as

---

54   Interview, December 5, 2005.

developed by Katz, 2004 and described earlier) and locate them on a continuum.[55] This continuum lays out the movement (or the development) between levels of political consciousness.

Many of the opportunities that residents create and discover in the space of the garden are the epitome of *resilience*. Residents discover in the gardens their own inventiveness that enhances their practices of survival and restores their dignity as meaningful participants of society. The idea of objectification of "species-being"—gardeners reinstating their connection with their environment (and with each other) through its production in their own image—is probably the most salient act of resilience. Those mostly marginalized groups of residents found a way to reconstitute their identity in the social realm. Through work of sheer inventiveness, people made the gardens into a space of comfort and balance, "oxygen," celebration, aesthetics, and so forth that dramatically reinforces their lives. The spatial gaze, stemming from the found inventiveness of the people, is a fuel and a resource in the praxis of survival. These practices of resilience are indeed, as Katz indicates, not only fortifying the actors and supporting their recuperation with the energy, pleasure, and balance they provide, but they are also self-reinforcing through their contribution to participants' sense of identity.

Moreover, these found opportunities (for resilience and survival) are only a stage in the development of the spatial gaze of residents. The newly found opportunities are constitutive of the further development of residents' critical thinking and political consciousness, political discourse, and political practices.

---

55 Katz, C. 2004. *Growing Up Global: Economic Restructuring and Children's Everyday Lives*. Minneapolis: University of Minnesota Press

PART II
The Spring of the Commons

Expanding the scope of discussion from the individual to the collective, the experience of community gardens also entails a social experience—and a very different type of social experience. As community gardeners, residents have a qualitatively different opportunity to engage in discourse and practices on a range of social issues such as environmental problems, redistribution rights, urban development, and more. Their experience of the garden has the potential to render them as urbanites who are truly engaged with their living environment, critically examining it, and capable of reacting to it and transforming it.

With *space* as the central coordinate (horizontal) by which other things are made possible, the *social* is a complementary coordinate (vertical) that activates various possibilities. The two coordinates together set up the nexus for discussing the social production of space. The discussion on the *collective* serves as the layer on which the transition towards something that is bigger than the individuals is allowed to happen, and demarcates the important interim stage between the level of the individual and the level of the institution.

Chapter 4
The Formation of the Collective

This chapter focuses on the *group* that produces community gardens: its formation, its boundaries, and most importantly the social visions and goals that the group develops, adopts, and adheres to. It specifically discusses the contributors and challenges that are present in the processes whereby a collective takes shape and place. These are processes and forces that are integral to the *formation* of the collective as a group, solidified by shared identity and goals. Then the *structure* of the collective is examined and its three structural units—garden groups, neighborhood coalitions, and New York City gardeners—are presented. The first unit, the garden group, is given the most attention as the main building block of the collective. The meaning and essence of the collective are dissected from various perspectives of gardeners, activists, and representatives of organizations; and the many negotiations on and challenges to the structure of the group are brought forward. The last segment—the *infrastructure*—reiterates the significance of space as the arena and breeding ground on which a formation of the essential organs of a collective can be developed.

The Formation of a Collective

While attending a plot in a garden is possibly a solitary activity, there is nothing private in being a member of a community garden; community gardens are produced collectively. Though the garden may not necessarily be inaugurated by a group, as will be discussed shortly, it is never, by definition and in practice, a one-person project. The collective of community gardeners is being cultivated from the ground up; it is produced and maintained through the practices of neighbors over space and their experience in it.

Residents collaborate for years in order to produce the garden; they co-determine and co-use the space of the garden as part of their everyday life experience. The real value and uniqueness of the gardens can be realized through their production by different people with different ideas, inputs, and needs. At the same time, through this shared production individuals become part of a shared product. Individuals certainly grow and develop in the process of producing the garden, but they are not growing independently of each other. Rather they learn together, socialize together, and become attached to a shared product and to its other producers. As Bell, a gardener from the Lower East Side, suggests:

[H]ere you interact with the people and there are many people with many different personalities; it is really like a neighborhood community, and it's everybody in a joint effort, and because everybody is caring for something that is just greater than himself, it is not just about having a plot and growing a few things, you know; it's actually making something beautiful for the whole community and sharing that with the whole community.[1]

The collective, then, evolves and develops through the process of collective production of space. This aspect of the collective is particularly important and singles out the struggle of community gardens from other social struggles. Other forces that bring residents together and consolidate them as a collective in a political struggle are also influential in the case of the gardens. But unlike other forms of urban movements—where the relationship of collectivity may be more instrumental or one-dimensional—the collective of gardeners is based upon a multi-dimensional relationship between participants that is based on an *ongoing* and **daily** co-production of space.

Since the early 1970s community gardens typically have been inaugurated by a few neighbors coming together sporadically to create a small group. Sometimes it was a single person who initiated the action and inspired others to follow suit. At first they planted flowers on the edge of a rubble-strewn vacant lot and then moved on to treat the whole lot, clearing it and planting vegetation. At different points during the process, as more residents joined the effort to change their neighborhood, they further established their garden group, organized its membership, and decided on its rules. They were driven to action by the condition of their immediate living environment; the solution was to alter the space in order to alleviate this condition. People had to learn from practice how to become resourceful and get things done. As one activist suggests, "in New York there wasn't that much sense of entitlement in the early 1970s. People just took over land and used all the resources that they could and scrounged materials and went out and tried to get things donated, or purchased things themselves."[2]

Rene from the Bronx tells the story of the inauguration of her garden—currently comprised of 40 members—in 1985 by one person:

Across the street from where I have lived they were supposed to build an additional three to five houses but when the developer got to that site he found that it was nothing but bedrock, so instead of building on that site single family homes he abandon[ed] the project and he left an empty lot. And so for a couple of years the lot became a place for garbage and debris and all sorts of things and it wasn't until I was looking at my window one day and I saw a guy with a shovel. His name was Jose Lupo and I asked what he was doing he said, "I'm

---

1   Interview, August 19, 2002, East Village, Manhattan.
2   Interview, November 8, 2005, Harlem, Manhattan.

going to clean up this lot to make a garden," and so I said, "Can I help you with it?" And he said "sure."[3]

Mike from the East Village, and a landscape designer by profession, also presents a one-person story about his garden (which was inaugurated in its present shape in 1995 but had previous stages), albeit in this story the leading person motivated others due to her leadership skills rather than actual actions:

> She was really amazing; she didn't do a lot of the physical work herself but she just had people from all over, architects, engineers, me and all these people that just got involved and built the grape arbor, we built the shed, the lawn, cleaned all the space up, took the fence down, opened it up. She started having regular meetings, got all these people involved. And it really started to shape up and that is when I started to get really excited about it.[4]

Green Thumb: Municipal Oversight of the Garden

By 1978 the city recognized and valued the effort of these groups, and established Operation Green Thumb to encourage the formation of other such groups.[5] Green Thumb provides licenses to all community gardens that are under municipal jurisdiction as well as encouraging the gardens of the Trust for Public Land (TPL) and the New York Restoration Project (NYRP) to be licensed as well and receive various services from the city. Licensing a garden group requires at least 10 listed members, a requirement which defines the basic formal unit of the collective.

The emphasis on the size of the collective is not irrelevant; it is an important aspect in the community gardens world, since numbers translate into money and power. From Green Thumb's perspective, numbers are the core element of administrative operation. A representative of Green Thumb explains how the budget is allocated based solely on the number of residents Green Thumb is serving:

> Once we were moved from the department of general services to the Parks [and Recreation] Department there were different expectation of us, I think the expectations of us are now higher. And our eligibility category for the money that we get from HUD [US Department of Housing and Development] which is where our money came from, it used to be acreage, acre-treated, and that was not a very accurate way to measure what was going on with gardens and also we were continuously losing land so we were not going to be able to meet our

---

3   Interview, November 2, 2006, Bronx.

4   Interview, October 5, 2006, East Village, Manhattan.

5   Francis, M., L. Cashdan, and L. Paxson. 1984. *Community Open Space: Greening Neighborhoods through Community Action and Land Conservation.* Washington, DC: Island Press.

goals. So we were able to change our eligibility category to public service so now how they measure out success is by how many people we work with not how many acres. So that changed a lot, it was in our interest to keep as many gardens as possible whether they were functional or not. Now it is in our interest to keep gardens that are functioning, functioning well, but there is no particular incentive for us to continue to preserve the land for potential gardening. If nobody is gardening there we do not have anything to do with that anymore. We get rid of it if we can; it is a liability for us. [...] So it makes a big difference. We can focus as much as we do on workshops and training and stuff because we changed that. It doesn't matter how much land we have. If there was one garden potentially we can still function.[6]

Therefore, the task of Green Thumb is to train the gardeners and help them solve their problems in order to keep their membership viable. But Green Thumb does not perceive protecting space for gardens as part of its goals; gardeners not gardens are important.

Presented in this way, Green Thumb is lacking a social or environmental agenda regarding the gardens: it is presented as a technocratic agency whose only mission is reaching its goals. Green Thumb provides help and services to gardeners, but at the same time allows the loss of potential land for gardens. This ambivalent behavior is explained to the gardeners in an appealing package: "We don't want to spend a lot of money recreating something that you created and building something that Parks creates—that is not a garden. Your sense of ownership is lost [...]."[7] Green Thumb, then, is interested in preserving the grassroots character of the gardens—something of great importance to the gardeners; but this attitude also makes it easy for Green Thumb and the Parks Department to lose garden land for other use.

Facing such an attitude, many gardeners define themselves as the "stepsons and daughters" of the city. They feel that the city cannot ignore them completely, but at the same time it does not encourage them and does not protect them with the same formal status and importance given to public parks and playgrounds. Green Thumb has a somewhat peculiar status within the city government. It cannot advocate for community gardens against the agenda of the administration, but it serves and supports the gardens that often challenge and contest the administration.[8] Understanding this partial role of Green Thumb further consolidates the collective of

---

6   Interview, November 15, 2006.

7   Jack Linn, Assistant Commissioner of the Parks and Recreation Department in response to the inquiries of gardeners in regard to unfulfilled promises of the department to invest in basic infrastructure, such as rundown sidewalks; April 22, 2006 at the Second Annual Gardeners' Forum.

8   Though stories suggest that at the height of the crisis in the late 1990s some figures from Green Thumb were trying to clandestinely organize the gardeners against Mayor Giuliani's efforts to eradicate the gardens.

gardeners as an independent group; they know that they have to look after their own interests and not rely on outside entities, such as Green Thumb, to represent them.

The Independent Collective of Gardeners

In the late 1980s gardens were losing legitimation that they hitherto enjoyed from the city, as their existence suddenly came into conflict with re-emerging market forces. This change in the state of affairs was probably *the* watershed moment for the commencement of an identified collective of gardeners in New York City. This moment pulled gardeners outside of their gardens, galvanized and united them in a struggle for the right to be and remain community gardeners. The collective was formed through a dialectical process in which the struggle for recognition and legitimization fostered the consolidation of individuals as a defined collective. Part of the struggle was in effect about asserting and receiving recognition of their *group* identity.[9]

Motivation of individuals to organize themselves around a certain interest and act together, to be produced as a collective, arise "despite the rhetoric of hypermobility" of people in the global era. It is because "most people still live very localized existences. [...] people still live and work and interact in very localized settings. [... and] some people, in effect, are more 'local' than others."[10] The significance of locality is that members of the community have a common investment in the same territory—in land. However, "investment" has conflicting meanings. On the one hand, there is the maximization of use value, where "communal living space" has meaning and use as a "community." On the other hand, investment entails the pursuit of exchange value, where the meaning and use of space is that of a "commodity." This duality can be developed into a more detailed "interests mosaic" of a given locality. The conflict that arises between the pursuit of use value and the pursuit of exchange value in the locality is that between interests of accommodation and interests of accumulation: usage, security, amenity, and autonomy versus exchange, equity, liquidity, and legacy.[11] The struggle between community gardeners and the city administration represents "competing discourses of rights" over urban space; property rights versus rights

9   Apfelbaum, E. 1979. "Relation of Domination and Movement for Liberation: An Analysis of Power between Groups." In W.G Austin and S. Worchel (eds). *The Social Psychology of Intergroup Relations*. Belmont, CA: Wadsworth, pp. 188–204. Here Apfelbaum refers mainly to racial minority groups' struggles, but the notion of struggle and resistance as they relate to a group's identity formation is applicable to nonracial conflicts.

10   DeFilippis, J. 2004. *Unmaking Goliath: Community Control in the Face of Global Capital*. New York: Routledge, p. 6.

11   Davis, E.J. 1991. *Contested Ground: Collective Action and the Urban Neighborhood*. Ithaca: Cornell University Press.

to public space.[12] Community gardens are a locally emplaced phenomenon; they are part of a neighborhood space and serve the residents of that specific locale. They became sites of conflict over investment (private versus public) that elicits the formation of the collective. The collective of gardens struggles over its rights for the land and the use value it affords for its members (against interests of accumulation).

Property rights articulate and distinguish the haves from the have-nots and the "abled" participants in the pursuit of the value of space from the "unabled" ones. "For property, of course, is the embodiment of alienation, an embodied alienation backed up by violence. More accurately, property rights are necessarily exclusive: the possession of a property right allows its possessor to exclude unwanted people from access."[13] Put more simply, private property in essence encloses the space from public access and usage, or at the very least denies the possibility to participate in its production according to the collective needs, image, and history of the public. This notion of exclusion suggests, in Fraser's more elaborate terms, that the formation of the collective of gardeners materialized as a struggle for both redistribution and recognition:[14] a redistribution of the values of space in a more egalitarian way (in particular for those members of society who traditionally do not have access to space as a commodity); and a struggle for recognition of a large group of "very localized" urban residents whose voices are traditionally not influential in matters of land use which directly and significantly affect their everyday life. For the most part community gardeners belong to a social group deprived of access to ownership of land. They are predominantly home renters rather than home owners and fall in the lower brackets of income distribution (see Chapter 1, Maps 1, 2 and 3).[15]

From the moment of its formation, the size of the collective was a very important component in the struggle over space. Activists of the New York City Community Gardens Coalition (NYCCGC), neighborhood coalitions, and supporting organizations of community gardens were working to create as large a collective as possible. They needed both a high number of gardens and a high number of gardeners in order to generate political clout and to protect and normalize this type of open space. The collective of community gardeners citywide needs to maintain its size, and expand, in order to thrive. The structure of the collective is influenced by this philosophy of maintaining a large, cohesive, and organized collective.

---

12   Staeheli, L.A., D. Mitchell, and K. Gibson. 2002. "Conflicting Rights to the City in New York's Community Gardens. *GeoJournal*, 58: 197–205.

13   Mitchell, D. 2003. *The Right to the City: Social Justice and the Fight for Public Space*. New York: The Guilford Press, p. 19.

14   Fraser, N. and A. Honneth. 2003. *Redistribution or Recognition? A Political-Philosophical Exchange*. New York: Verso.

15   As gentrification of inner-city neighborhoods progresses and expedites the peripheralization of the urban poor, the socio-economic composition of the collective is changing.

Structure

The collective of community gardeners is dynamic in structure; it redefines itself over time and with changing external forces.[16] We can demarcate three units as comprising the structure of the collective of gardeners in New York City: garden groups, neighborhood coalitions, and the citywide body of gardeners.

Garden Groups

The basic unit of the collective is the garden group which increasingly, due to encouragement from various supporting organizations, follows variations of a certain internal structure. The garden group usually elects a coordinator or an executive director, a deputy coordinator, a treasurer who manages the garden's budget (income from membership fees and donations, and expenses for tools, plants, parties, reconstruction), and several more elected members to serve on a board. In some garden groups each board member is in charge of a committee—such as landscape, events, membership, and fundraising. The garden group, organized by the board, holds monthly general meetings for all members of the garden, and separate meetings for the board and the committees. This general structure that varies among gardens is defined in the bylaws that also outline the format of elections to the board of directors and other general rules regarding the operation of the garden.

This structure is often negotiated and challenged within the collective and, as will be shown shortly, has implications for the sustainability of the collective. For example, community gardens in the East Village in Manhattan with a predominantly white membership (mostly of artist newcomers to the neighborhood in its early stage of gentrification) demonstrate a much more developed and democratic structure of board and rules of operation than those of the Puerto Rican gardens in that neighborhood.[17] In addition, there are still claims voiced by gardeners criticizing some garden groups whose operational structure strongly deviates from the ideal model of operation. Nevertheless, it is now safe to say that

---

16 Urban social movements and collective actions are often comprised of a group that defines and redefines itself dynamically in response to processes and forces that surround and affect it. See for example the struggle of the housing movement in NYC (Lawson, R. and M. Naison. 1986. *The Tenant Movement in New York City, 1904–1984*. New Brunswick: Rutgers University Press; Marcuse, P. 1999. "Housing Movement in the USA." *Housing, Theory and Society* 16(2): 67–86); the open space movement in NYC (Site, W. 2003. *Remaking New York: Primitive Globalization and the Politics of Urban Community*. Minneapolis: University of Minnesota Press).

17 Martinez, M. 2002. "The Struggle for the Gardens: Puerto Ricans, Redevelopment, and the Negotiation of Difference in a Changing Community." Unpublished dissertation, New York University.

the participatory and decentralized structure characterizes most gardens citywide, regardless of the ethnicity and prior experience of their members.

Having the opportunity to exercise so much control over the environment is not only positive; it also generates internal conflicts that challenge the structure of the garden group. There are conflicts over the vision for the garden and the right way to materialize it, conflicts over usage of the space, its aesthetics, methods of creating it, and so forth. Other problems pertain to the centralization of control and power in the hands of a few while alienating the others, or the opposite problem of many "free-riders" who enjoy the hard work of the few who assume responsibility and control. The opportunity to be involved, to control, and to determine is a tricky affair and can be translated very easily into some sort of oppressive power. The negotiations that garden groups as well as the broader collective maintain over the structure of the group are intended to avoid this slippery slope.

Still, there are gardens citywide that are controlled by one or two figures who act as dictatorial leaders of the garden and centralize all the power in their hands. In many of these cases, the "dictators" are the people who established the garden, who become possessive about it, and refuse to allow "intervention" in its management. Gardens like this are characterized by a small membership, usually of close friends and family of the leader. There is a sort of forgiveness among gardeners towards those, usually elderly, leaders who manage the garden as their own. The approach towards most such leading figures is to wait until they step down and then change and democratize the structure. A representative of the Trust for Public Land in New York City had to deal with several such one-leader gardens in the pool of 67 gardens that it purchased. The situation of a "dictatorial" leader is very delicate, since those leaders are usually much appreciated for their contribution to the community (they established the garden), but at the same time are resented for not sharing control with others. As the TPL representative stresses:

> [...] there has to be a leadership *group*, there has to be a plan for new people coming in and for succession, and it has to be a *group*. You just can't have an individual, you just can't accept that if someone is making it impossible and you can't change it, you may just have to wait until they leave the neighborhood or pass away, god forbid, because you can't necessarily fight with them in the streets for years. It can literally happen if you try too hard to displace them, but it is not a viable model. All of our dictatorial, single leaders that we still have in some of our gardens are a problem. There are not many of them left because we managed, by a combination of the surrounding community and that dictatorial leader, to make some changes. But the more dictatorial leadership can be really hard, it includes people who most of the world thinks they are angels, and they may be, but they make it impossible for anybody else to take any active role in the garden.[18]

---

18   Interview, October 17, 2006.

The drive towards emplacement of a participatory model as the structure of the garden group is mainly because it is necessary for the survival of the collective and for the production of the garden as a certain kind of social platform.

One aspect of this social platform is a diverse and open environment where arguments and quarrels are part and parcel of reaching agreements and collaboration. Though socially challenging, this environment offers a learning experience for some gardeners. However, for others, these challenges can be negative, as Claudia from East Harlem suggests: "[...] for the people participating who don't see the benefits they get as worth the politics and the personal disagreements about petty things, they tend to burn out. And if there are no new people who want to step up into their role, then what do you do?"[19]

It is easier to avoid having a highly participatory environment and some gardens definitely do not take up the challenge. For example, there are gardens that function more like social clubs and definitely do not serve the general public. It is possible to deduce the ideal vision for the gardens from the bad examples that exist. Rene, a gardener and activist from the Bronx, offers examples of gardens that significantly diverge from the ideal:

> One thing you have to realize is that there are good gardens and there are not-good gardens. People say "We have to talk about all the gardens" but face it, there are some gardens that shouldn't be gardens. Number one, if you don't allow public access. Number two, if you have it for your personal use. Number three, if you allow cars in the gardens, to park in the garden, you allow gambling, you allow playing numbers, you allow drugs in the garden, illegal sale of alcohol and cigarettes. Some gardens are like that, I'm just laying it out. And the Garden of Paradise and the Clinton Garden [two gardens in the neighborhood] they were like that, because it became like a club house. They weren't really gardening in there. They had their little plant on the side but basically it was a club house. It was like a social club, exclusive.[20]

Community gardens can become exclusionary space, much like private property, when people centralize control over this public asset. Some people take advantage for their personal gains, thereby threatening to delegitimize the gardens for everyone else. Therefore, many activists like Rene and Claudia emphasize the importance of the public-ness of gardens and act to root out those that fall short of this basic standard. Past experience showed that not only is this form of community gardening not desirable (ideally), it is also not *sustainable*. In the struggle for preserving the gardens, the "social club gardens" saw the highest number of lost spaces due to their isolation from their own community and the community of gardeners.

After the struggle for preservation, many garden groups worked to establish a more sustainable structure. TPL, for example, decided to purchase gardens from

---

19 Interview, November 8, 2006, Harlem, Manhattan.
20 Interview, November 2, 2006, the Bronx.

the city in 1999 and save mostly those endangered plots that exhibited a developed and participatory organizational structure. A representative of the organization reviews TPL's logic at the time:

> The gardens we chose were the more active of the gardens that were for auction. It was not the TPL's philosophy to buy land and be sort of a land bank, land that *might* at some point be run by these communities. We wanted to acquire gardens that had some sense of organization in operating them. So we chose ones that were more or less working and that the leaders were quite active and we knew, it was our vision anyway, that as much as feasible the decisions about community gardens future should be made by the garden volunteers. That is the way community gardens emerged in NYC. Community gardens didn't start based on an agency starting them off. They grew up from the grassroots and so we did have a philosophy that they should continue to be grassroots run.[21]

The collective of community gardeners likes to define itself as inclusionary, participatory, and self-sufficient. Therefore, exercising control in an exclusionary manner disconnects groups from the collective. But in the same vein, assuming too little control also hinders the participatory model of community gardens. Conducting monthly meetings, for example, is a challenge in many gardens due to weak participation. Gardeners across the city describe the phenomenon of the "core 10"—the group of gardeners who always attend the meetings, workdays, and events—while other garden members are much less committed to contributing to the collective effort. Again, Claudia from East Harlem describes the situation in her garden:

> I see members of my garden and every time they see me they come and complain to me about things that bother them or things they would like to do in the garden, "Why don't we do this." And I say to them, "Great, come to the meeting and we will talk about it at the meeting," and then they don't come to meetings. So for me, if you can't make the effort to come to the meeting it must not be such a big deal for you [...]. But that is not the way it works. I don't just go and do something about it. It has to be something that is decided upon in the meeting with everyone who wants to be involved in making decisions farther. And our bylaws specifically say any major decision about the garden, permanent and semi-permanent structures and all that stuff, have to be made at the meeting.[22]

Claudia suggests that lacking the responsibility to show up for the meetings is indicative of indifference. But her cynicism is probably related to the frustration experienced by those (core 10) who invest a lot of effort in meetings and discussions compared with those who do not. The relatively flexible structure of

---

21   Interview, October 17, 2006.
22   Interview, November 8, 2005, Harlem, Manhattan.

the garden groups allows for varying degrees of member participation, but the differential contribution by different participants may cause both frustration and operational difficulties. For example, decisions are made by a smaller group and are later challenged by those who were not part of the decision making; processes of decision making and execution of decisions are slower and often require modifications (sometimes after the fact); frustration is experienced by both sides, and sometimes members get so upset that they leave. Nevertheless, having a loose organizational structure, with varying degrees of participation, is alluring in many other ways and opens up multiple possibilities.

Defining Membership

Community gardens encompass opportunities for different people to be part of the garden group in order to satisfy different needs in different ways. Those needs and the ways to fulfill them are constantly negotiated. Negotiations also revolve around questions such as: Should people who are not interested in gardening become members? What about people with very little time to work in the garden? Can members rarely come to meetings and workdays? Most garden groups try to find flexible solutions to these dilemmas in order to be as communal as possible. They introduce minimal requirements for membership and are not very vigilant in keeping to them.

Discussions about these issues reveal the effort to define the collective of gardeners. They suggest that the collective is both a means to protect and sustain the gardens and an end in and of itself: a collective of people who realize the opportunities of the space without overexploiting it. Mike from the East Village describes an ongoing negotiation at his garden in regard to a segment of the collective—"the mothers":

> Mothers are here in the day a lot and they may not be active in the garden but it is still so important that they are here using it. I've always argued, we have people in our board that are very active physically who complain that some mothers don't pull their weight maintenance wise, and I'm always going back to the fact that I'm just glad they are here and using the space the way we envisioned it: being used which was big time for families. And I don't have family so […] and the people that are more actively involved are usually single, or they are couples but without kids. Those are the people who are more involved, show up for the workdays more often.[23]

According to Mike, as well as other gardeners, the contribution to the common wealth that defines someone as part of the collective varies. It may be using the space, enjoying it, and thereby making it into what it is supposed to be—a space to accommodate the needs of the community in many different ways.

23   Interview, October 5, 2005, East Village, Manhattan.

The demographics of the membership of gardens also present some challenges to the current boundaries of the collective and its future. Negotiations over defining the collective go beyond the level and form of participation, to the characteristics of its population. The aging population of gardeners is one major issue that activists and organizations are addressing with great attention. According to activists and supporting organizations, the majority of gardeners in New York City are over 50 years of age, and in some neighborhoods the average age is higher. Therefore, one of the biggest challenges they face is bringing the younger generation to the gardens and maintaining their stability as the older generation relinquishes the responsibilities of operating the gardens. A representative from the Green Guerillas explains:

> A lot of gardens that the Green Guerillas work with, their gardeners are seniors. They are not going to be able to do the stuff they are doing for much longer. And if there are not younger people coming in, they are not going to look so good and the city is not going to be happy about keeping these gardens.[24]

Green Thumb perceives neglected gardens as a liability and is quick to define them as inactive and transfer them to its Housing Preservation and Development department (HPD) or to some other city agency. If the collective will not stretch its boundaries to include enough younger people who can replace the older ones it is most likely that the space of those gardens will be lost forever. Many gardeners from gardens with a predominantly elderly membership understand this problem and try to act on it. Rene from the Garden of Happiness in the Bronx explains that gardens in her area of the Bronx are trying to bring the younger generation to the garden in many different ways. These include summer camps, environmental education programs in the garden, and youth participation in operating the farmers' market. But the issue requires more attention than individual gardens reaching out to a younger membership. Neighborhood coalitions, the citywide coalition, Green Guerillas, and other organizations employ various mechanisms to prevent gardens from becoming derelict and to help garden groups with their efforts to bring in new members.

The Garden Group and the Community

The garden group unit does not function as a dispersed spatial enclave that offers an alternative urban experience; nor is it spatially ascetic. It is connected "vertically" to other units of the collective as well as "horizontally" to its immediate community. The collective depends on the approval and acceptance of the surrounding community, and especially on those in the community who are not personally engaged with the garden. That is, garden groups have to find ways to become integrated into the rest of the community, since an alienated community— one that does not support the garden and its activities, and which does not feel

---

24   Interview, November 8, 2005.

part of it—is not sustainable for the garden. Threats to a garden group from an alienated surrounding environment vary from concrete acts of harassment such as complaints to the police when events are taking place in the garden, through acts of vandalism and the lack of political support in the community board decision during an Urban Land Use Review Procedure.

Mr. Thomas, a gardener from East New York in Brooklyn, is talking about the new garden he and fellow members had to create in a relocation site, several blocks away from their homes, after their old garden was taken away for development:

> And you have it today a blossoming upcoming, thriving, lovely space that in the beginning when we first came here the community, especially nearby people, did not embrace us because whenever you're new you have to prove yourself::Who are you? What do you want? Why are you messing up [...]? So the gates were burglarized four different times [...]. But over time now we are part of the community. We are definitely an integral part. You may see it while you are here, it is kind of early but as people walk by you may see them stopping and holding on to the fence and asking questions or whatever.[25]

When well integrated into the community, gardeners experience insignificant amounts of vandalism and theft. Toys for children are left there, cherished as collective property for everyone to use; plants with their fruits and flower are respected when in individual plots and generously shared when in the collective plots. Tools, sculptures, and personal decorations are kept intact despite frequent visits by nonmembers to many of the gardens.

Neighborhood Coalitions

At times, the collective grows beyond the basic unit of the garden group to the community level and forms a neighborhood coalition. There are several neighborhood coalitions that were active at the time of writing: Harlem United Gardens (HUG), Bronx United Gardens, La Familia Verde Coalition (in the Crotona neighborhood, Bronx), East New York Garden Association, and Greenpoint-Williamsport Gardens Coalition in Brooklyn. There were many other such neighborhood coalitions, especially in the heyday of the struggle (from about 1998 to 2002), but many disappeared after the struggle had subsided. These coalitions, that bring together from as few as five to as many as 60 gardens, try to follow the same structure as the basic unit—a board of directors, coordinator(s), and committees.

The most important purpose of the neighborhood coalitions was to provide a platform for the organization of the struggle to save the gardens. Since gardens were threatened in every neighborhood, it helped to have a coalition acting both on the local level with the local politicians and community boards, and on the citywide

---

25 Interview, May 6, 2006, East New York, Brooklyn.

level to raise public awareness and galvanize the public, green organizations, and the city council in favor of the gardens. Rene tells how La Familia Verde gardens coalition was established in the Crotona neighborhood. The story suggests how the threat to the gardens brought people together to form a group as well as articulating the importance and value of community gardens to the neighborhood for themselves and for the greater community:

> Back in 1998, that was when the Giuliani Administration started [to] auction off gardens; everybody was scrambling to find out which gardens were up for auction, which gardens are going to be taken away in their neighborhood. So I'm in community board 6, there were about four to five gardens that were on the auction list. But the thing that was so unique about our community board is that our community board is pro open space, pro gardens. So, it was very important at the time to first of all galvanize those gardeners that were on that list, but also to galvanize those gardeners that were not on the list. Because you can't lie on your back and say, "I'm safe now." No, because just like they felt they were safe, the same thing can happen to you the next day. So it is important to show unity and strength and that all the gardens in the area start sending out pamphlets: let's meet together, let's form a coalition, let's fight this because we can't fight it alone, let's fight it as a group. And that is what happened. Green Guerillas was very instrumental in helping us in terms of organizing it; not organizing it but helping us in terms of strategizing. First of all we got our meeting place, and then we talked about what we are going to do, and then we came with a name, how we are going to call ourselves, and then we came with a mission statement, and then we had weekly meetings finding out what is going on. And then as a group, as a La Familia Verde group, we went to the community board and said this is who we are; this is what we have done to the people in the community; this is how we contributed to the progress of things; this is what we have done to help our community. What can you do to help us?[26]

Neighborhood coalitions serve as a link between the local garden groups and the "global" citywide struggle. The need for this link diminished to some degree when the settlement was signed in 2002 between the city and the gardens. However, many neighborhood coalitions kept on going and currently serve as a supportive network to the gardens, connecting the gardens with the various supporting organizations, and dealing with local problems. Marvin, a coordinator in the Harlem coalition, explains how the settlement changed the dynamic of his neighborhood coalition:

> What I found out is that during the struggle it was important for everybody to work together and everybody was united. But with the legislation those persons who got their garden secured, they couldn't care less about everybody else. So what happened, our coalition was about 25 to 30 different gardens and we had in

---

26   Interview, November 2, 2006, Crotona, the Bronx.

every meeting about 20 gardens represented, maybe 30 people in every meeting and since the legislation we have maybe 10 [in the meetings and] about maybe 12 [gardens as part of the coalition].[27]

The Harlem United Gardens (HUG) coalition is still very active with networking, fundraising, and organizing social events for the gardens in East Harlem. In 2007 having this functioning body proved crucial when Harlem gardens were threatened again. HUG (with the help of other organizations, such as More Gardens! and Time's Up!) orchestrated a publicly covered struggle over the 20-year-old Nueva Esperanza community garden. Gardeners challenged the approval and financial support ($12 million) that the city handed to the plan to develop the site of Nueva Esperanza on East 110 Street for a museum of African art. They argued that the museum was a pacifier for the local community and a cover-up for the 116 luxury units in the tower on the corner of Central Park above the museum. In the last months of 2007 and the beginning of 2008 gardeners and activists were encamping full-time in the garden and using other methods to raise public and political support for Nueva Esperanza and the other endangered gardens in East Harlem.

Several organizations, most notably More Gardens! and Green Guerillas, are working closely with neighborhood coalitions citywide, helping them solve problems pertaining to their neighborhoods and further facilitate their organization. Green Guerillas, for instance, has been helping the now thriving East New York Gardens Association (representing 50 gardens in the Brooklyn neighborhood). Recently, Green Guerillas established a branch in Bedford-Stuyvesant, Brooklyn, a neighborhood abundant with community gardens that is going through intensified gentrification and demographic changes. More Gardens! was located in the South Bronx and worked closely with the Bronx United Gardens coalition to develop a plan that will be more sustainable for gardens in the course of the neighborhood development. In 2006–07 More Gardens! relocated its base to East Harlem to work together with HUG to struggle for the over 20 endangered gardens within a 1-mile radius.

The contemporary role of coalitions in the neighborhoods is further discussed in Part III as their actions are usually more focused on branching out of the realm of the garden(s) to include other community organizations and to propel what I refer to as the *institutionalization* of community gardens.

The Citywide Body of Gardeners

The next unit of the collective is the body of gardeners in New York City. This unit is much more dispersed than the previous two and therefore harder to demarcate. The New York City Community Gardens Coalition (NYCCGC) represents the most salient effort to clearly establish a representative unit of gardens citywide. It was established in 1996 and has operated since then with changing degrees of

---

27 Interview, November 22, 2004, Harlem, Manhattan.

intensity and representation (its missions, activities, and the challenges it faces are discussed in Part III). Apart from NYCCGC, gardeners citywide also gather for Green Thumb activities, most notably the various workshops and the annual GrowTogether event. A representative of Green Thumb explains that one goal of GrowTogether is to instill a sense of being a collective among gardeners:

> We just sent out a big mailing to all the gardens where they haven't been to any workshop this year. Because I think it is very important that they do at least something so they meet the other gardeners. [...] But we try to have the gardeners feel like they are a group; have them maintain that feeling of being a group. And I think that the GrowTogether is probably the biggest part of that. When you see the gardeners at the beginning of that, at the opening session when there is 1,000 people in there, they definitely feel like they are a group, they are all into it. Of course it is not everybody but 1,000 people is a lot. That is more than one person for each garden coming. It mostly is gardeners; it is not like half greening professionals.[28]

In this respect, the observation that the collective of gardens follows a "flexible organizational pattern and that coalitions are structured on tenuous relations among gardeners, bound by individual commitments"[29] is correct but insufficient. It ignores the forces that are at play in propelling this collective and securing its survival. The flexible structure of the collective is most evident in the context of three coalesced components: 1) urban processes; 2) the scope of the collective's vision; 3) the strategies of operation. The collective is most united and mobilized when facing a colossal threat to its existence: it acts to protect the gardens from destruction. There was one such climactic point in 1999 when gardeners from the entire city were united in fighting the Giuliani Administration's plan to auction off many community gardens. The effect of this threat was so pervasive that some activists perceived Mayor Giuliani's plan as the most crucial moment in the formation and consolidation of the gardens movement in New York City.

In an attempt to weave together the many parts that comprise the development of political practices and discourse, the multilayered connections that individuals and collectives establish simultaneously towards space and towards each other seem to be crucial. Articulating these connections helps us understand the types of relations that mobilize people into action and develop their critical thinking. There is an accumulative force that stems from personal and collective gains and needs; a force that not only entails a strong commitment to the collective, but that also develops a certain life perspective and group identity. The time and energy that gardeners devote to their environment stretch beyond gardening to deal with neighborhood affairs and city politics. These experiences shape them further as

---

28   Interview, November 15, 2006.
29   Hassell, M. 2002. *The Struggle for Eden: Community Gardens in New York City.* Westport: Bergin & Garvey, p. 64.

a collective with specific practices, discourse, and identity. The creation of such a solid body of community gardeners bolsters the collective's ongoing fortitude beyond the commitment of individuals to it at a certain point in time.

Space as Infrastructure for Collective Formation

The collective of gardeners in New York City is dealing with many challenges of defining and sustaining itself. Crucial for its sustainable future is the capacity of the collective to find long-lasting solutions to internal organizational problems as well as to further facilitate a more participatory and decentralized model of operation within the different units that comprise it.

This chapter steps beyond the level of individuals' interaction with space to explain the formation of gardeners as a group and its collective action over space. It lays out a descriptive infrastructure of the gardens movement in New York City that will be utilized for discussing broader urban issues in the following chapter. Many individuals of this collective share a marginalized position in the city and are excluded from certain services and from positions of power. These individuals are working together to redress their marginalized position through collective spatial practices in the garden. Though this collective presents a diversity of cultures, ethnicities, races, social class, and background it is connected through certain praxis and through a developing consciousness. Claudia from Green Guerillas shrewdly observes one component that the collective of gardeners share:

> It is really inspiring for me to work with community gardeners because I think they are the type of people, they are all so different, different ages, different backgrounds, different interests, different gardening skills but the thing that they all have in common is that they are all people who are very aware of the immediate physical environment that they live in. And they are motivated enough so they don't just accept the world as it is. They don't just go home from work and they are tired and they close the door and watch TV. They come home from work and they look out on their block and they [are] like, "What is going on right here where I live and how can I make it better; how can I do this?" And that is really inspiring to be with people who feel motivated and inspired to make a change in their immediate local environment and actually go out and do it. And have the imagination to see what could actually be done in a space that doesn't really look that good. And then to actually pick up a shovel. I just wanted to say that I find them very inspirational people to work with and they are people who have an incredible amount of imagination and they are not afraid to make an effort to realize the dream that they have if it is just in terms of the physical space or the community that comes together around the gardens, it is beautiful and inspiring to work with those people.[30]

---

30   Interview, November 8, 2005.

The collective of gardeners is formed first and foremost through its actions over and through *space*: the appropriation of space for gardens and the protection of space as gardens. These are the quintessential purposes of the core units of the collective (that is, garden groups, neighborhood coalitions, and NYCCGC) as well as other supporting organizations. Moreover, all the individuals who participate in community gardens, in any type of activity, contribute to this collective action— whether they are aware of it or not. This is to say that on a very basic level community gardening is a *collective spatial practice*.

Space therefore is the bedrock on which this collective is formed and according to which it establishes its parts and develops its various organs. Space is the very basic infrastructure for the collective to evolve and develop its spatial practices and spatial understanding. This spatial collective action entails experiences, practices, and interactions that are in varying degrees of dissonance and disagreement with the dominant space and its (social) structures. The space that is collectively produced contradicts the common prescription and perception of space—constituted by the dominant spatial logic—as it complicates, reveals, and challenges it. It is because of this specific infrastructure, the argument goes, that community gardening as a spatial collective action or social production of space[31] offers a sort of de-veiling experience, whereby hegemonic prescription is unraveled by the sight and experience of the alternative that was formed. The spatial practices that result in new structures, experiences, meanings, and understandings are transformed, almost inevitably, into political practices. This transformation of spatial practice into political practice is the quintessence of the power of space according to Lefebvre and his successors. The production of spatial diversity (differential space, alternative spaces, spaces of hope) is the underlying mechanism for new and alternative political practices.

Some of the components of the gardens' spatial diversity are manifested through the reconstruction of past landscapes, celebration and aesthetic experiences, and ownership and control over space. These are experienced predominantly as individuals' opportunities. But social production of space through which space is produced and experienced collectively and which, at the same time, diversifies the general urban landscape shapes the social and spatial collective with new and alternative quality.

---

31   Lefebvre, H. 1991. *The Production of Space*. Oxford: Blackwell.

Chapter 5
Actually Existing Commons

So far community gardens have been discussed in terms of introducing alternatives to residents who tend them: space that negates the dominant space and, within its boundaries, offers an alternative life experience alongside an alternative understanding of the world. As such, community gardens can be usefully perceived as heterotopias.[1] The gardens offer alternative aesthetic experiences that challenge prevailing aesthetics and maxims regarding the authority to decree aesthetic judgments; they bring nature to a place that is dominated by the lack of it. Moreover, the environmental practices and discourse that are being cultivated in the gardens often revolve around organic growing, addressing the need of the community for fresh produce and fighting pollution.[2] The gardens also allow for the creation of an alternative social experience that challenges the prevailing alienation of people from their physical and social living space and the general insensibility towards underprivileged groups. It is not only about a space to meet members of the community or to practice gardening, but also a space to become significantly involved in developing the very idea of what a city should be and a place to nurture this idea. The gardens offer an alternative psychological experience by enhancing people's sense of control, empowerment, and belonging. Last but not least, the gardens are a collective spatial product. They are about taking part in the production of space and determining the spirit of the city.

This chapter intends to identify manifestations of an alternative political project through the examination of community gardens as heterotopias in the form of *actually existing commons*. By doing so it also revises and updates the notion of the commons in the modern urban space. The commons are a way of thinking and operating in the world, a way of organizing social relations and resources. Actually existing commons, just like actually existing neoliberalism, have multiple modalities, mechanisms of development, and "diverse socio-political effects."[3] Actually existing commons are live relics of the ideal of the commons; they are never complete and perfect and may even have components that contradict the

---

1   Harvey, D. 2000. *Spaces of Hope*. Berkeley: University of California Press; Lefebvre, H. 2003. *The Urban Revolution*. Minneapolis: University of Minnesota Press. See the introduction for the discussion on this concept.

2   Hassell, M. 2002. *The Struggle for Eden: Community Gardens in New York City*. Westport: Bergin & Garvey. Hassell presents an overview of the role of community gardens around the world in introducing urban agriculture.

3   Benner, N. and N. Theodor. 2002. "Cities and the Geographies of 'Actually Existing Neoliberalism.'" *Antipode* 34(3): 349–79, p. 353.

ideal type. Nevertheless, even in the face of pervasive neoliberal ideology and practices, alternatives *do exist* and they pave the road to new politics and another possible world.[4]

There are urban systems that might be considered as actually existing commons, such as the collective ownership of housing designated for and managed by poor urban populations in the form of limited equity cooperatives,[5] or workers' cooperatives that act as a common resource of livelihood.[6] In addition, for many years the streets were used as public spaces following the sense of the commons. They were used by various groups for multiple purposes before their commodification and transformation into a spectacle as part of the general conceptualization of the urban "in and through the 'bureaucratic society of controlled consumption.'"[7] Community gardens in New York City are paradigms of counter-hegemonic spaces. They were produced collectively by residents of the most neglected locales, only to later become targets for capitalist development. These attempts to enclose the gardens initiated a counter-reaction by local residents.

The conceptualization of the commons as property with no rights allocation and regulation and as belonging to everybody, and hence to nobody,[8] underlined the debate over the commons for the last six decades. But this debate goes far back, to the commons in England that sustained the livelihood of landless serfs. The English commons were criticized already in the fourteenth century as obstacles to more *productive* forms of agriculture and were eventually enclosed in the eighteenth century.[9] Half a millennium later Hardin's (1968) "tragedy of the commons" continues this thread of thought, suggesting that both population growth and the motor of human behavior—of productivity and competitiveness, and of maximization of short-term individual gains—make a property which is everybody's and nobody's unsustainable. This predicament could be resolved through the allocation of property rights: either nationalizing or privatizing common resources.[10]

---

4   De Angelis, M. 2003. "Reflections on Alternatives, Commons and Communities or Building a New World from the Bottom Up." *The Commoner* 6, p. 2.

5   Saegert, S. and L. Benitez. 2005. "Limited Equity Housing Cooperatives: Defining a Niche in the Low Income Housing Market." *Journal of Planning Literature* 19(4): 427–39.

6   DeFilippis J. 2004. *Unmaking Goliath: Community Control in the Face of Global Capital*. New York: Routledge.

7   Lefebvre, 2003, p. 4. In many neighborhoods of New York City, streets and sidewalks still function more like the commons or are periodically closed to create temporary commons.

8   Gordon (1954) in Mansfield, B. 2004. "Neoliberalism in the Oceans: 'Rationalization,' Property Rights, and the Commons Question." *Geoforum* 35(3): 313–26.

9   Goldman, M. 1997. "'Customs in Common': The Epistemic World of the Commons Scholars." *Theory and Society* 26(1): 1–37.

10   Hardin, G. 1968. "The Tragedy of the Commons." *Science* 162: 1243–8.

A critical response to Hardin suggests that the real tragedy of the commons became the tragedy of their disappearance. Nationalization or privatization of the commons entails their enclosure, and in effect eliminates the complex systems of self-regulation that local people constructed over many years of devising sustainable ways of using common resources.[11] It is the neoliberal practices of commercialization that destroy the commons[12] and these practices of enclosure continue to serve as a generative force for capital expansion (not only as an historical precondition for the development of capitalism).[13]

In the late-modern city ideas of the commons as well as "public space" changed their meaning. Since the 1970s, as the logic of capital and private interests took over more and more urban space, the city has undergone an intensive erosion of spaces that were considered public and that served public purposes: streets, parks, and squares.[14] The rationale for this massive restructuring of the city through privatization was the need to address problems of crime, drugs, and general neglect that became prevalent in the 1970s and 1980s. The type of solution that was adopted, though, was not necessarily directed to benefit the public—and in many cases resulted in the exclusion and displacement of local residents. But the pervasive privatization of "the public" initiated a counter-reaction of groups such as the squatters' movement and the "reclaim the streets" and "critical mass" movements, struggling to reclaim space that was once designated to serve the common wealth and that was now in the service of private, capital accumulation interests. The 2011 protests against neoliberal capitalism (in Europe and the US) took a dominant character of the right to the city movement and recreated temporary commons in city centers.

While the contradictory political ideologies of capitalism and socialism both propel a property regime designating space to be either private or public, respectively, the commons are neither of these.[15] In order to understand the commons the neoliberal control over the knowledge that determines the workings of society and ultimately "the realm of what is defined as the commons" should

---

11 Monbiot, G. 1994. "The Tragedy of Enclosure." *Scientific American* 27(1): 159. See for example Mansfield (2004), which gives his analysis of various privatization and nationalization mechanisms that were deployed over fisheries in the North Pacific Ocean.

12 Hardt, M. and A. Negri. 2004. *Multitude*. New York: Penguin.

13 De Angelis, M. 2007. *The Beginning of History: Value Struggles and Global Capital*. London: Pluto Press. The enclosure act of the eighteenth century is considered the epitome of primitive accumulation that enabled and propelled capitalism (as well as urbanization).

14 Harvey, D. 2006b. *Spaces of Global Capitalism: Towards a Theory of Uneven Geographical Development*. New York: Verso. Times Square is a public space designed as a tourist attraction. See also Sorkin, M. 1992. *Variations on a Theme Park: The New American City and the End of Public Space*. New York: Hill and Wang.

15 Hardt, M. and A. Negri. 2009. *Commonwealth*. Cambridge, MA: Harvard University Press.

be exposed.[16] This requires moving away from thinking of the commons only as a material and finite resource (to be either freely consumed or regulated from overconsumption), and towards reconstructing the prevailing ownership model in a way that accentuates the multiple and contradicting possibilities embedded in *property*.[17] Part of this is the need to start re-envisioning the commons outside of the public–private dichotomy and introduce the social, cultural, and political practices that allow new possibilities in order to reconstitute the commons as an object of thought. The commons can then serve as a platform for envisioning and developing an alternative framework for social relations and social practices.[18]

The urban commons follow several core characteristics. First, the urban commons are produced. Second, they offer a set of livelihood qualities over which rights are negotiated: dwelling, open space, recreational and social space, movement in space, and control over space, to name just a few. Third, the urban commons fulfill these and other social needs in a non-commodified manner. Fourth, they require communities to operate them through collaboration, cooperation, and communication rather than through private interest and competition. Altogether, the commons provide the opportunity "to obtain social wealth and to organize social production."[19] In what follows, the space of community gardens is deconstructed into its Lefebvrian constitutive elements—material, representations, and lived—thereby exposing the potentialities of this space as actually existing commons. By doing so I wish also to enable further examination of other actually existing commons with an elaborated theoretical framework that includes new discourses, practices, lived experiences, social relations, and subjectivities that are contingent on the commons.

Material/Absolute Space: Actually Existing Space of the Commons

Material space is an actual space of fixed, identified, and discrete entities. It is a space of experiences and practices and is therefore defined by its use-value—its non-commodified and non-commercialized qualities.[20] It is the actual space of the garden with the soil, plants, animals, and people. In New York City today this material space amounts to about 650 community gardens, which add up to roughly 90 acres.[21] It was already suggested that community gardens in the United

---

16   Goldman 1997, p. 3.

17   Blomley, N. 2004. *Unsettling the City: Urban Land and the Politics of Property.* New York: Routledge.

18   Hardt, and Negri 2009.; De Angelis 2003.

19   De Angelis 2003, p. 6.

20   Harvey 2006b; Lefebvre, H. 1991. *The Production of Space*. Oxford: Blackwell.

21   As unlikely as it might sound, the city's database of community gardens does not provide information on acreage. Green Thumb's fund allocation is based on the number of participants rather than the size of their property.

States (re)produce the space of the commons, most notably because gardens are communally and locally managed and enable some self-sufficiency for their participants. Existing accounts of community gardens as the commons emphasize the materiality of the commons—that is, the actual space and its actual usage. The gardens offer some material resources such as land, air, healthy food, community space, and "land-based enterprises such as cooperative market."[22] To this list we can add recreational and cultural facilities.

The collective of gardeners perceive this material space as their own due to the strong psychological ownership that gardeners derive from the shared use of a common space (as Chapter 3 showed). Control and (collective) determination of space are among the components that were reviewed as crucial for the development of this psychological ownership (or sense of ownership) as such. The very idea of communal authority of space challenges contemporary common sense. Hence, protecting the space of the gardens requires almost perennial struggle, new conceptualizations, and legal solutions. For years community gardens had no legal status; they were considered vacant lots in municipal procedures of urban planning. The assault on New York City's community gardens, led by the Giuliani Administration in the late 1990s, blatantly defined the gardens as an irrelevant phenomenon belonging to a bygone era, and as one that should therefore be uprooted to make way for progress and development. Due to the massive public outcry, the Administration failed in eradicating the gardens, and three different schemes for protecting the space from privatization were put in place. None of them offers a permanent solution but all three are means to sustain the material space of community gardens in the neoliberal city.

In the first scheme, about 400 gardens were preserved under the Parks and Recreation Department (PRD) of New York City. According to the city's law, the land under this jurisdiction cannot be "de-parked" without a very complicated process of approval involving also state-level intervention and if a garden is taken away, a similar-sized piece of land must be offered to compensate for it. Another reassurance for preserving the gardens is the informal commitment of the municipality based on their *modus operandi*. "If it walks like a duck and it quacks like a duck, it is a duck ... [the] commitment on the part of the City [...] is just as strong as your commitment to gardening. If you continue gardening we protect your garden."[23] Thus the Assistant Parks Commissioner assured gardeners advocating for an official community gardens policy. This informal commitment suggests that as long as there is a community that maintains the garden it will be protected. However, a history of the destruction of beautiful, well-maintained gardens by the city makes this statement, which is not anchored in policy, questionable.

In a second scheme, 67 gardens were purchased by the national nonprofit organization Trust for Public Land (TPL) and preserved as land trusts. Members of these gardens are entitled to collective legal ownership over the space as

---

22   Linn, K. 1999. "Reclaiming the Sacred Commons." *New Village* 1: 42–9, p. 43.

23   Jack Linn, April 22, 2006—the Second Annual Gardeners' Forum.

long as the gardens are maintained as an inclusive community resource.[24] By transferring the gardens to land trusts, TPL ensures that the property is taken out of the market system, albeit not permanently: ongoing community participation in the production of gardens as spaces that serve the commonwealth conditions their sustainability. The model of community gardens that is developed by TPL emphasizes the organizational infrastructure that is needed to sustain the space: if the gardens are underused and do not operate as a collective resource, or are no longer needed and appreciated by the community, they lose their purpose and their right to be preserved. This model most closely takes after the ideal of the commons as serving a wide variety of public purposes and needs through communal authority and maintenance. It also reaffirms the idea that the commons necessitate a community—the commoners—that holds the authority to manage the gardens. For the commons to succeed, a TPL representative asserts, "it is crucial that there will be an organization that can run [the gardens] for public benefit [...] Community gardeners will play a significant role in governing the organizations and the gardens will become increasingly important for their neighborhoods as a result of being as public as possible."[25]

The third scheme for sustaining the material space of the gardens was proposed by the New York Restoration Project (NYRP), a nonprofit organization that purchased 59 gardens in order to expropriate the land from the market. Unlike TPL, NYRP emphasizes land rescue over community participation and runs the gardens with a vision of maintaining them as enduring beautiful green spaces. To bring this vision to life, NYRP hired professional designers who redesigned each of the gardens. In some of the gardens the community was successfully integrated after the fact, but in many gardens the community remains alienated from the space—which was not produced by them and according to their needs and vision. These gardens require paid staff members to regularly maintain them (similarly to urban parks). It is therefore difficult to evaluate the number of NYRP gardens that can be thought of as instances of the commons any more than public parks. Although the organization protects these spaces from the market, many of the gardens fail to serve the needs of the community and are perceived by many residents and activists as uninviting, elitist spaces. In addition, the centralized management of these community spaces by NYRP makes their ongoing existence overly dependent on the organization's funding.

A representative of the Council on the Environment of New York City (CENYC) unravels the main challenges of collective ownership and the possibility of a long-term existence of the commons:

---

24   The Manhattan Land Trust and the Bronx Land Trust received the entitlement of the gardens in 2011. The Brooklyn-Queens Land Trust was in the last stages of getting the entitlement at the time of writing.

25   Interview, December 5, 2005.

So now, for instance, you have a garden that is physically permanent and you got a small group of gardeners gardening there; that group doesn't recruit new people, they get old, they move away, they die, they can't garden and now you have a piece of land that is permanent but who is taking care of it. So I think that those kinds of issues, being prepared for those kinds of things and having some kind of answers to those questions, I think is important.

It is suggested, then, that it is not enough to fix the common resource in place since long-lasting commons depend on mechanisms that could also sustain the mode and spirit of collective production of space.

These are the three solutions that were set up between 1999 and 2002 to protect and secure community gardens as a common resource in New York City. Although each of these solutions contains caveats that challenge the future status of community gardens, none of the three falls strictly within the private or public definitions of urban space and can inspire our rethinking of alternative constellations of the urban commons. These schemes emphasize the coexistence of the local-material space *and* a collective—a community that maintains the space collectively—as the two key coordinates which make community gardens commons.

The intervention of civil society via nongovernmental organizations (TPL and NYRP) in protecting the commons reflects not only the eroding capacity of state and local governments to protect public space in the face of economic pressure to privatize it, but also—as the different interpretations given to the gardens by these two organizations highlight—provides some initial insights regarding who holds the power to control and define the commons. Nevertheless, space-centralists such as Lefebvre and Harvey emphasize the prominence of the material space in the evolution of any alternative set to transform the dominant social structure. They insist that real and meaningful alternatives could only flourish from a collective action rooted in the reworking of the material space.[26] The gardens are actually existing spaces, present in absolute space and time. The collective actions of gardeners are aimed at protecting and controlling the material space. However, protecting the material space is not enough; it must be intertwined with mechanisms of cooperation and communication that activate the community of users, produce alternative knowledge, and offer alternative experiences of space. The other two facets of space—lived and representations—reveal the mechanisms that not only produce the material space but also change the meaning and value of its materiality.

Lived/Relational Space: Gardens as Carriers of Culture

Lived space is space as experienced through images and symbols which do not submit to quantifiable rules. It is the emotional quality that is exerted from space—emotional values and meanings which are immaterial but objective. It is the realm

---

26   Lefebvre 1991; Harvey 2006b.

of collective memories, cultural symbols, and personal history. "As a space of 'subjects' rather than of calculations, as a representational space, it has an origin, and that origin is childhood, with its hardships, its achievements, and its lacks."[27]

The lived facet of the space of community gardens has multiple expressions in images, memories, emotions, identity, and everyday practice. As most gardeners are external or internal immigrants to New York City, the gardens are experienced as symbolizing the landscapes of childhood which they left behind. With those past landscapes are also various practices that get re-enacted in the gardens, propelling a strong place attachment and identification with the living environment as well as a sense of ownership and control (as discussed in Part I). The space of the gardens is an important common resource for creating meaning and enhancing a positive emotional experience of the living environment.

The most physically salient aspect of the symbolic meaning of the gardens is their constitution as carriers of cultures within the city. The hegemonic culture expresses itself in space, deploying mechanisms that marginalize the expressions of other cultures. In spite of this, the space of the gardens is reappropriated and used to celebrate these silenced cultures. In this context culture refers to a groups' differentiated manifestation of social and spatial practices. Culture is seen as a *boundary of difference* that is socially constructed by power relations.[28]

Most of the gardeners in New York City define themselves as Latino (coming mostly from Puerto Rico) or as African American (first or second generation in the city arriving from the rural South). Since gardens are very local sites that are established by members of the surrounding building blocks, ethnic segregation, or even ghettoization,[29] explains why many of the gardens are clustered as single-ethnicity gardens. Community gardens that are located in more ethnically diverse neighborhoods reflect this diversity in their membership. With the gentrification of inner-city neighborhoods that had left no Italians in Little Italy and displaced Latinos from Loisaida (now the East Village) and Spanish Harlem, there are also growing numbers of white members in community gardens.

One of the signature characteristics of the pool of community gardens is their variety. Each garden allows for a uniquely different experience of space with its own arrangement, aesthetics, usage, and colors. This diversity is possible because gardens are spatial expressions of a specific group that was not formally trained in urban planning or landscape architecture and that does not attempt to implement principles from these disciplines. This enables gardeners to express

---

27   Lefebvre 1991, p. 362.

28   Appadurai, A. 1996. *Modernity at Large*. Minneapolis: University of Minnesota Press.

29   Thabit, W. 2003. *How East New York Became a Ghetto*. New York: New York University Press. Thabit defines ghettoization as the process in which various policies prevent minority populations from living in white neighborhoods and force them into neighborhoods that are slated for minority occupancy in which deprivation of infrastructure maintenance, policing, and education reifies their marginality.

and experience their culture collectively (rather than privately, in the confines of their own homes). And indeed, various aspects of culture are realized in the gardens through a rich experience that engages aesthetic and culinary preferences, rituals, customs, artistic expressions, and social interactions. While presenting an impressive diversity, gardens could be roughly divided into three types: the casita gardens, the farm gardens, and the eclectic culture gardens.

Casita gardens are predominantly Latino in population, and are typified by the *casita*—literally a "small house" in Spanish—that "imitates traditional rural Puerto Rican homes, [the design of which] has been traced back to the indigenous Tainos ... [it is] brightly painted to evoke dwellings on the island."[30] The casitas are used to store food and musical equipment for cultural celebrations, and serve as a cozy seating place for the gardeners. Latino gardeners generally perceive the garden as important mostly for community development and as a space for social and cultural gathering over preservation of open space and civic agriculture.[31]

One of Manhattan's East Village gardens—9th Street Community Garden and Park[32]—that was able to preserve its Latino membership despite gentrification is also one of the most beautiful and well-maintained gardens in the area. Stretching over one-eighth of a block it displays flowers, plants, and trees along brick paths that branch through small, intimate and greenery-surrounded seating areas. There are several collective food production areas and some individual plots. The area in front of the casita is especially large, with some shaded picnic tables and other seats and several big barbeque grills. Walking in the garden in summer evenings one can observe the members gather around the tables, dining together, listening to live Latino music that is frequently performed there during the gardening season (see Figure 5.1).

In some casita gardens, gardeners realized the strength of the garden as a space for cultural transmission and officially transformed it into a cultural center. A successful example is the Rincon Criollo (literally Creole Corner) Cultural Center[33] in the South Bronx. Ilya, a gardener from the Bronx, describes it like this:

> Now we have Rincon Criollo that is very amazing. It is basically a cultural center/ music center. So they have enhanced their own Bomba and Plena which are the music and dance of Puerto Rico which are dying out there, but are very fresh over

---

30   Martinez, M. 2002. "The Struggle for the Gardens: Puerto Ricans, Redevelopment, and the Negotiation of Difference in a Changing Community." Unpublished dissertation, New York University, p. 67.

31   Saldivar-Tanaka, L. and M.E. Krasny. 2004. "Culturing Community Development, Neighborhood Open Space, and Civic Agriculture: The Case of Latino Community Gardens in New York City." *Agriculture and Human Values* 21: 399–412.

32   http://www.youtube.com/watch?v=HX2aoxY8ZvM.

33   http://www.centroculturalrinconcriollo.org.

Figure 5.1   9th Street Community Garden and Park, East Village, Manhattan

here. And the groups are using it, people all over the world know that … so that's very large. And their community garden is very open and everybody can come in.[34]

The effort to revive Puerto Rican working-class dances (the Bomba and Plena) within the context of the casita garden is eloquently explained in the *NY Latino Journal*:

For Puerto Ricans, whose immigrant experience has been one of displacement rather than assimilation, the creation of casitas like the one at Rincón Criollo,

_____

34   Interview, May 31, 2005, the Bronx.

has enabled us to take control of our immediate environment and, in the process, to rediscover and reconnect with our cultural heritage.[35]

The second type of garden, "farm gardens," are predominantly African American in population, and their space is organized mainly for food production (Figure 5.2). African American gardeners manifest their culture in the very practice of gardening and the level of self-sufficiency that it provides. In most cases farm gardens are community oriented, though somewhat differently from the casita gardens. Many gardens organize food giveaways and community feed-ins several times during the gardening season. Another form of community engagement is weekly distributions of donated food and fresh produce to the community's poor. As another example, the Euclid 500 Block Association community garden in East New York, Brooklyn, placed a basketball hoop at its rear to attract adolescents from the streets into the garden—exchanging gardening hours for permission to play. Some farm gardens also organize free workshops on food canning, knitting, papier mâché hat-making, and so forth.

Figure 5.2   Farm garden: East New York Historic Garden, Brooklyn

---

35   Carlos "Tato" Torres. *NY Latino Journal*. retrieved September 2006 from http://nylatinojournal.com/home/culture_education/ny_region/rincon_criollo_more_than_just_a_little_house_in_the_south_bronx.html.

Figure 5.3   Eclectic culture gardens, Manhattan

Most prominent in farm and casita gardens is the cultivation of vegetables and herbs that are part of the ethnic cuisine but either unavailable or unaffordable. For the African American kitchen, farm gardens produce many leafy vegetables such as collard and kale, and a variety of corn and tomatoes. Casita gardens are known for their hot and sweet peppers and various herbs. Thomas from East New York talks about this aspect of gardens as carriers and educators of culture:

> There are multi-cultures in this neighborhood and so some of the gardeners plant specifically for people's personal [use] … so I get to hear about things like kulolo and all these different things that I would never hear about if I would […] go to the supermarket to get my produce, I wouldn't know about all of this.[36]

The third type of garden, the "eclectic culture garden," is characterized by a predominantly white membership and is located mostly in areas that have undergone or are undergoing gentrification. Membership in these gardens is generally younger than in the other two types. Eclectic culture gardens usually present a mixture of social space and gardening space with more areas of plant display. This difference is probably related to the higher socio-economic status of the gardeners: unlike in the other two types of gardens, here food production is less of a necessity. Eclectic culture gardens in the Lower East Side, Manhattan— an area that has faced intensive gentrification since the mid-1980s—are better connected to various green and neighborhood organizations that support the gardens than the casita gardens in the same neighborhood. They therefore have more resources to invest in the design of the garden and in the quantity and type of events that they offer.[37]

As the name suggests, eclectic gardens feature a variety of cultures. Underlying many of these cultures is a sensibility that stretches from environmentalism to paganism.[38] The annual Earth Day festival and the biannual solstice event, celebrated in these gardens, are among the festivities that manifest these sensibilities. In the event calendar of these gardens one would find yoga and tai chi classes, lectures on nature, eclectic music performances, and movie screenings (see Figure 5.3).

It is interesting to note that the first historical phase of community gardens in the USA was a government-initiated poor relief program (inaugurated in 1894 in Detroit). Designed with cultural assimilation in mind, the program served, according to its proponents, as a melting pot in which new immigrants would assume an industrious persona and learn the American way. The program was widely adopted in the USA because of its financial success as a "welfare to work" program but faded away after the Second World War.[39] In contrast, the contemporary phase of

---

36 Interview, May 6, 2006, East New York, Brooklyn.
37 Martinez 2002.
38 Hassell 2002.
39 Bassett, T. 1979. "Vacant Lot Cultivation: Community Gardening in America, 1893–1978." Unpublished manuscript, Department of Geography, University of California,

community gardens reflects an opposite approach to cultural assimilation. While the mechanism of a melting pot de facto aims at ironing out differences and assimilation into the hegemonic culture, in their current phase community gardens celebrate past experiences and revive cultural practices rather than repressing them.

Community gardens as the commons offer a daily and direct experience of a multiplicity of cultures, expressed in the physical environment and the social practices that are engraved into the landscape of the city. Beyond this opportunity to voice suppressed cultures, they afford and actualize a lived experience of space that emphasizes diversity, celebration, aesthetic expressions, attachment and belonging, and connection to collective and individual history. Understood through the lens of lived space, the gardens support the ongoing production of a community of residents and afford an alternative lived experience within the modern urban environment: by integrating historical and cultural experiences into daily lives, this lived experience de-alienates the physical and social environments of the city.[40]

New Framework, New Discourse: Representations of Space

Representations of space are dependent on "the frame of reference of the observer."[41] These are the abstract perceptions of space that are determined by or relative to the tools and frameworks used to formulate them. Representations of space belong mainly to the realm of knowledge (*savoir*), where understanding is interwoven with ideology and power. Neoliberal representations of space are produced by the scientific gaze of planners, engineers, and urbanites. These representations revolve around the exchange value of space—its quantifiable and commodifiable qualities allowing capitalism to progress by effecting how we perceive space (as a commodity) and changing its materiality.[42]

Representations of space are the result of cognitive acts: schemes, ideas, and understandings forming a body of knowledge that is imbued in formal (that is, education) and informal (i.e. culture/media, common sense) ways. I understand the production of knowledge in community gardens as a social activity that

Berkeley; Lawson, L. 2005. *City Bountiful: A Century of Community Gardening in America*. Berkeley: University of California Press.

40   Lefebvre, H. 2002. *Critique of Everyday Life: Foundations for a Sociology of the Everyday* (vol. 2). New York: Verso. Lefebvre sees everyday life as the terrain for a movement to grow and transform the prevailing system. For example, he is fascinated with festivals as an act of production of space for creative cultural celebrations that bring to life (albeit temporarily) the histories and cultures of people.

41   Harvey 2006b, p. 122.

42   Ibid.; Lefebvre 1991.

encompasses both learning and communicating ideas about the world.[43] Representations of the gardens do not follow the neoliberal rationale. They develop as the gardeners learn "to see together, exchange their feelings, values, categories, memories, hopes and observations as they go about their everyday affairs."[44] The gardens are sites where local knowledge—the knowledge of a multiplicity of groups that is unique to them and is created in the everyday context of their lives—surfaces and becomes conscious and voiced.[45]

There are various ways by which the exchange of memories, values, feelings, and daily practices allow for knowledge to develop, percolate, and deep-root itself among its producers in community gardens. Some of these mechanisms are formal, while others are informal and spontaneous. This knowledge is practical and skill-based as well as discursive and abstract. It is available for the wide public (in both formal and informal modes of learning) and considered here as another layer of the commons that community gardens enable.

Negotiating the Production of Knowledge

Some of the new knowledge is produced by the various support organizations and set out top-down on the gardeners. There are established, formal ways to learn about community gardens, gardening, the environment, and more. The most obvious and popular one is Green Thumb's workshop program, which offers workshops on various subjects each month in four of the city's boroughs. These workshops include such topics as organic gardening, compost mastering, winter gardening, rainwater harvesting, building basic raised beds, tree pruning, companion planting, herbal wellness, garden design, chicken coop building, bird house building, extending the growing season, conflict resolution and basic communication skills, dividing perennials, creating a rain garden, and so on. Some of the workshops are delivered by outside professionals and some by the gardeners themselves. Gardeners are welcomed and encouraged to participate in as many workshops as they like. But there is an additional incentive to participate in these workshops beyond knowledge and skills acquisition: in each workshop Green Thumb distributes some necessary garden supply (for example, tools, soil, compost, seeds, etc.).

The workshops serve as a resource for knowledge and at the same time as a means for Green Thumb to establish more control over the gardens and to tighten the gardens' dependency on the organization. Gardeners cannot just ask Green Thumb for supplies since some supplies are conditional on participation in a workshop. This top-down production of knowledge (and the dependency it

---

43 Following Vygotsky, L.S. 1978. *Mind in Society: The Development of Higher Psychological Processes*. Cambridge, MA: Harvard University Press.

44 Lynch (1976) in Pile, S. 1996. *The Body and the City: Psychoanalysis, Space and Subjectivity*. New York: Routledge, p. 24.

45 Sandercock, L. 1998. *Towards Cosmopolis*. London: Wiley.

induces) raised a counter-reaction from gardeners protecting their own organic mechanism of production of knowledge. Gardeners, Claudia from Harlem reports, argued: "I have been gardening for 20 years, why do I have to go to workshops on gardening in order to get supply, I already know what I'm doing."[46] Green Thumb responded to these voices of resentment by diversifying the workshop program and having "some of the workshops on very basic information and some of the workshops unusual, interesting information that an experienced gardener will not be bored to go to."[47]

There are interesting negotiations over the production of knowledge, even when this knowledge is practical and skill-based. As a city agency, Green Thumb represents the dominant representation of space: the space of city planners, architects, and designers, and of their tools—maps and guidelines. Gardeners, on the other hand, produce their own representations of space—their body of knowledge—out of their own experiences of space and interactions with it (but not entirely detached from abstract knowledge). One recent example of negotiations over top-down and bottom-up production of knowledge comes from the city's new ordinance regarding the illegality of all of the structures that exist in community gardens, most notably the structure of the casitas. This means that Green Thumb will have to issue dismantling orders that no doubt will cause frustration and antagonism among gardeners. Green Thumb's solution is to approach the subject from the existing channels of knowledge by introducing workshops that are described as follows:

> [...] ever wondered, "What can I build in my community garden?" Green Thumb has recently been issued guidelines from the Department of Buildings for acceptable structures in community gardens. Come hear the director of Green Thumb talk about how these guidelines will impact existing and future gazebos, casitas, shade pavilions and sheds [...].[48]

Green Thumb attracts gardeners to this workshop by offering hammers, work gloves, and the guidelines themselves. Therefore, the collective knowledge of building and gardening that was developed through the years and guided the production of various structures and landscape is now being challenged by top-down, unified, guidelines. This authoritative knowledge, at least for the time being, is infiltrating not as a decree but through the chains of dependency that were established between gardeners and Green Thumb over more recent years. It exemplifies the duality between organically developing, spontaneous knowledge and top-down, directive knowledge or between dominant representations of space and alternative, grassroots representations.

---

46   Interview, November 8, 2005, Harlem, Manhattan.
47   Ibid.
48   Green Thumb's Fall 2007 Workshops Calendar.

Bottom-up Production and Sharing of Knowledge

There are also formal ways of producing practical knowledge which are orchestrated by gardeners in community gardens. These include free workshops, lectures, afternoon programs, and summer camps. In 2007, for example, gardeners reported that 42 percent of the gardens were working with schools in the neighborhood, teaching students about plants, animals, and gardening.[49] Some gardens operate afternoon programs and summer camps and are involved with sustainable food programs in which young people are taught about locally produced fresh fruits and vegetables and are encouraged to eat and purchase them in farmers' markets. La Familia Verde gardens coalition in the Bronx offers several venues of environmental education for young people. Rene from the coalition describes them:

> We invited high school students and the project consisted of six different groups and of course our group dealt with gardening, teaching them the essence of gardening, making them aware where food comes from, asking questions, reading labels in a supermarket: "If you can't pronounce it why do you eat it?" Making them ask questions, having them taste fruits and vegetables that are grown in the garden […]. So educating people and local is the key. Sometimes we encourage young kids, who come to our summer day camp, to ask us questions about stuff and we give them fruits to take with them so they could sit down and discuss how it taste, because vegetables that you get from a farmers' market are so much different than what you get from a store, and compare quality and compare taste. Every other week we have a cooking demonstration at the farmers' market and that was provided through partnership with Just Food and the Health Department.[50]

Other gardens have programs for women and young people, facilitating environmental awareness and neighborhood empowerment and offering some skills to improve their competence in dealing with a challenging urban environment. Sam, a gardener from Harlem and a long-time activist, provides another example of an educational program run by his garden—Project Harmony:

> [The program is about] working with women and young people, with young people primarily, to learn about the environment, to do work based around the environment, to give them some hands-on things that they can do about their community, about cleaning up their community, working in their community, etc. So we try to teach them as much as we can about environmental issues and teach them how to work together and to solve problems and to create projects that they could do.[51]

---

49   Based on a self-report survey of 114 gardens conducted in 2007.
50   Interview, November 2, 2006, South Bronx.
51   Interview, June 8, 2006, Harlem, Manhattan.

These programs, then, offer young people and adults skills and a sense of competence and at the same time offer an alternative set of representations, starting from questioning the prevailing ones and proceeding to rethinking the place and role of residents and gardens within the urban environment.

In addition to formal learning programs, the gardens are also constituted as an informal urban resource for learning: they serve as a forum for a variety of spontaneous learning that is facilitated by the ongoing interaction with nature and the people that tend to it. One example is the "play gardens" where children can incorporate natural elements (such as sand, water, twigs) into their play while interacting with people of various ages who work in the garden—as an alternative to a secluded, age-designated, gated playground.[52] Mike from the East Village describes his garden's plans to better utilize the space for both formal and spontaneous learning opportunities:

> [The garden is] also an incredibly valuable resource, [our garden] is not being used as such so much, but for education, for kids and the schools. You see the schoolyards in New York, they don't have green, they don't have grass, they don't have trees. Kids don't get to see what plants are like or where food comes from. I think that that is hugely valuable and we need to massage that a little more and get more students, groups, and classes. We used to have summer camps here and I want to do that more. It is also valuable for adults, I learned a lot about plants getting involved here. [...] I look forward to this year as we are going to redo a lot of the wild flowers areas and the woodland garden and redo the lawn and actually get that done on a professional level to make it more of the kids' educational garden. Because how many places in the city are [there] where you can see wild flowers and native plants? There are few but not many so I think it will be a valuable resource for teachers.[53]

More subtly, spontaneous production and sharing of knowledge occurs in those daily unplanned interactions in the garden. In this quality of the environment lies one of the most important values of gardens—though incommensurable: it represents another aspect of gardens as a common resource. A representative of Green Thumb provides an example:

> [...] by being involved in the garden, [gardeners] learn a lot of skills that they can translate to the world outside of the garden. I think a lot of the new gardeners are Mexican and they don't speak English at all and when they get involved with the older gardeners who were there and they end up, the older gardeners end up doing a lot of things for them. Translating bills and telling them what they need to do. It is just a way, a really good way to mix people together in a way that

---

52   Hart, R. 2002. Containing Children: Some Lessons on Planning for Play from New York City. *Environment and Urbanization* 14(2): 135–48.

53   Interview October 5, 2006, East Village, Manhattan.

I think is actually significant rather than [...] I don't know what would be an insignificant way to mix people together. But to bring people together for a rally that is great; they had a great time but would they be connected after that? The garden, you have to go there pretty much every day and all the time, at least you are going to pass by all the time.[54]

These words capture the three significant potentialities of spaces like community gardens: the *diverse* collective *cooperating and communicating* to produce a collective resource in an *everyday setting*. The relative absence of spaces that are safe and open enough for such spontaneous learning and sharing of skills and knowledge underscores the uniqueness and importance of community gardens as the commons.

Gardeners appreciate the contribution of the gardens to their educational and personal development through attaining practical knowledge such as gardening and language, social, and leadership skills, as well as abstract knowledge that can generally be categorized as urban cognition. Mike from the East Village describes his learning curve in the garden, referring to gardening and organizational and leadership skills:

I did the lawn, I planted the wild flowers and I had a great learning experience, and the other thing was that at the time I really didn't know anything about plants [...] so this gave me a kind of a little laboratory to experiment with different plants. [...] After [the garden was] preserved we started restructuring the board and the membership. We had a member who used to be a corporate consultant, taskmaster. He was helping us get ourselves together; helping us just be more organized as a board and run more efficiently. [...] Now that it was no longer about battling development which was our focus, you didn't really have to think about running things in an organized way. So I was into that, I was into how it was changing and I was learning from it because at the same time I was building my business. So I was learning how to manage people. So it seemed to work well.[55]

Another aspect of gardens as a common urban resource is that they serve as "a little laboratory to experiment" and master different things such as planting, building structures, and organizing.

In addition to practical knowledge, the collective production of space propels the development of the socio-spatial understanding of residents that amounts to a conceptual framework regarding space and its users. This knowledge, as the following reveals, is not detached from practical knowledge or from practices. Rather, practical knowledge and practices contribute to the discourse that emerges. The collective production of space propels the emergence of the socio-spatial gaze

---

54 Interview, November 15, 2006.
55 Interview, October 5, 2006, East Village, Manhattan.

of residents that encompasses certain understandings and consciousness. The space of the commons allows for an alternative experience of everyday life which clashes with the dominant experience. As a result, a new consciousness is developed.

For some gardeners, the space of the garden and its practices served as means to articulate an understanding they held in some way before they joined the garden. Other gardeners found the space and its practices to be a platform and an opportunity for the unearthing and development of a new consciousness.[56] The scope of the (new) gaze, the alternative representations of space, can be analytically dissected into three thematic fields: urban and community political economy; environmental sustainability; and construction of class. Both practical and discursive bodies of knowledge serve as important collective resources for the gardeners, and both, as we shall see, are crucial for the reproduction of the space as the commons.

Urban and Community Political Economy

The discourse on urban and community political economy exposes the vision held by gardeners regarding the kinds of social relations and spatial practices they would like to accentuate in their urban life, alternative visions which are constructive elements of the commons. New knowledge and understanding regarding the meaning and practicalities of *community* are evident in the discourse of gardeners. Gardeners acknowledge the contribution of gardens to the safety and beautification of neighborhoods, and to social cohesion and social capital of communities.[57] But in addition, gardeners develop new representations of space that rely on critical examinations of notions of neighborhoods, communities, and the city, uncovering their unjust and uneven development.

Claudia from Harlem alludes to critical ideas regarding community and urban life, such as those of Jane Jacobs,[58] when asserting that "neighborhoods that got gardens in them are safer because there are more eyes on the street."[59] Most notably, when compared to other public spaces available in urban neighborhoods, gardeners identify the uniqueness of the space of the garden: an interactive public space used for a wide variety of activities such as gardening, barbeques,

---

56   Depending on their initial standpoint as described in Chapter 3—the "proactive" and "reactive" narratives of gardeners.

57   As research has indeed shown: see, for example, Hancock, T. 2001. "People, Partnerships and Human Progress: Building Community Capital." *Health Promotion International* 16(3): 275–80; Kingsley, J. and M. Townsend. 2006. "'Dig In' to social Capital: Community Gardens as Mechanisms for Growing Urban Social Connectedness." *Urban Policy and Research* 24(4): 525–37; Schmelzkopf, K. 1995. "Urban Community Gardens as a Contested Space." *Geographical Review* 85(3): 364–81.

58   Jacobs, J. 1961. *The Death and Life of Great American Cities*. New York: Random House.

59   Interview, November 8, 2005, Harlem, Manhattan.

meetings, and other social events, art and performance, and for getting to know your neighbors.

The gardens offer a certain experience that has been eroded from modern urban life, an experience that was overpowered by the importance of hyperspace, work, and well-orchestrated spectacles. The alternative vision of community is not a nostalgic longing for past days; rather, it was molded by present experiences that take place concurrently with the dominant ones. The gardens negotiate this dominance of experiences by putting forth an alternative daily experience of a strong and supporting community.

Ilya from the Bronx thinks within the paradigm of participatory and organically grown community as he presents the strength of community gardens in his neighborhood:

> [The] garden does that naturally; it grows, people plant things, we have sculptures, we got performances, kids grow up in it. And so it is all the people who are the community, sharing their vision of what they believe and what they want ... so when that happened and these communities came about, a community of strong nurturing people came about. And obviously the agency thought it is easily structured but the roots of the people are stronger than the roots of the trees we had planted and they were not going anywhere. [...] To me it is always the best school when the community is a part of where they learn from each other, create a place with the new comers and the old folks and help each other.[60]

Bell, a gardener from the Lower East Side, explains what connects people and defines them as a community:

> [H]ere you interact with the people and there are many people with many different personalities; it is really like a neighborhood community, and it's everybody in a joint effort, and because everybody is caring for something that is just greater than himself, it is not just about having a plot and growing a few things, you know, it's actually making something beautiful for the whole community and sharing that with the whole community.[61]

This discussion presents an alternative vision of community and encompasses a critique on the condition of community in the contemporary neoliberal city. According to this new meaning, residents produce a community not because they share a common characteristic, such as living environment, belief, or profession. Rather, they are a community because they cooperate, collaborate, and communicate on the usage, production, and maintenance of a common resource.[62] Rather than accepting the prevailing modality of competition and self-interest that inevitably

---

60 Interview, May 31, 2005, South Bronx.
61 Interview, August 19, 2002, East Village, Manhattan.
62 De Angelis 2003.

leads to atomization and hollows out the essence of community, gardeners are facilitating a new modality and definition of community that enhances their social cohesion, level of autonomy, and the intensity of social bonds.

Mike from the East Village presents another idea of how to use community gardens as a resource for sharing alternative knowledge in a more institutionalized way: "[using the gardens], the big ones, the very important ones, where someone is there a lot so that people could visit them and they will have a curator that can tell them about their history and their importance to the neighborhood. Look at that from an urban planning standpoint if you wanted."[63] According to Mike, then, an alternative narrative of the history of urban and community development can be presented to the public in the gardens through the story of the gardens themselves. The story itself has, of course, a crucial role in shaping the residents' sense of community.

The visions and crystallizations of an alternative community begin with a critical discourse on the uneven investment in neighborhoods and the uneven powers of communities. Sam from Harlem is somewhat reluctant to squarely define people as powerless, but underlying his description of the role of gardens in the community is the realization that the city is unevenly densified, and that some residents might not have the same privileges in urban public parks:

> [The gardens] are really important in this neighborhood because for the most part these are really dense neighborhoods. And as you can feel when you're walking up the street, the cars and the asphalt really make it hot in the summer time. You need some place that you can come out to. If possible, to have a sitting area that is right in your neighborhood, that is a plus. Not everybody gets out to parks and stuff like that. They might have a problem walking or might not feel safe in the bigger parks so they can have access to a garden or a small park in their community.[64]

According to these new representations of space, the production of space and the definition of community are tightly interlocked: the community is not merely a group of people who occupy a designated environment and operate within it according to its established purpose (such as clubs, religious institutions, parks) or even one that exhibits some resistance practices towards that environment (à la de Certeau).[65] Rather, it is a group that participates in the production of its own material environment according to its own culture, history, desire, and vision— and is thus constructed as a community.

Much of the knowledge about the city's political economy is produced and circulated as a result of the struggle, which has been waged since 1999, to preserve

---

63   Interview, October 5, 2005, East Village, Manhattan.
64   Interview, June 8, 2006, Harlem, Manhattan.
65   De Certeau, M. 1988. *The Practice of Everyday Life*. Berkeley: University of California Press.

the gardens. New representations of space were produced by garden groups and coalitions that fought for their gardens at community board meetings; mobilized support from the community and politicians; worked with lawyers to challenge decisions in courts; and joined protests, rallies, and demonstrations. A new body of knowledge emerges as gardeners realize their own (lesser) position within the urban power structure and processes of development.[66] Billy from Brooklyn explains their realization sarcastically:

> What happened to the idea of green space? Why take away the community gardens when the city owns the lots? Because they are poor neighborhoods, they don't deserve parks [laughing]. That is what Giuliani was basically saying by taking away the community gardens: "These neighborhoods don't deserve green space; they deserve crap housing, and still more crap housing."[67]

Gardeners realized that their neighborhoods are overly dense, and have the least open space per capita in the city and fewer public amenities than better-off neighborhoods. They realized that despite their contribution to their neighborhoods they are perceived by the municipality as a menace. They also learned that public or private investments are not intended to improve their conditions but actually further marginalize them. Gardeners protest about the over-ghettoization of their neighborhoods resulting from the overconcentration of public and senior housing and rehabilitation centers in their neighborhoods; the lack of groceries, schools, and parks; and the gentrification fueled by the municipality and the local growth machine that threatens their gardens and themselves with displacement.

Claudia from Green Guerillas talks about this production of knowledge as a process of "dis-naivete" through which the gardeners realized the real mechanisms that underlie the political structures: "I think those gardeners are pretty savvy and they know too that [the] Parks [Department] could arrange some kind of official signing of papers and transfer them back [for development]."[68] A representative from the Council on the Environment of New York City refers to it as "collective wisdom," crucial to the future of gardens:

> We also have all this experience of the struggle that we had for those years with Giuliani that people know what needs to be done. [...] But whether or not there is an official policy in New York City, I think that there is enough collective wisdom and knowledge and experience of the gardeners themselves and of all the greening groups and other people and founders too, people who supported

---

66 Laclau, E. and C. Mouffe. 2001. *Hegemony and Socialist Strategy: Towards a Radical Democratic Politics*. New York: Verso. The authors define antagonism, understanding of one's marginalization, as the first step towards awareness and the articulation of needs and demands.

67 Interview, July 29, 2006, Crown Heights, Brooklyn.

68 Interview, November 8, 2005, Harlem, Manhattan.

> community gardens over the years monetarily, and as long as there is an ongoing dialog about where we going, you know we don't stop talking about that. I think that is the key—to communicate. As long as communication continues to go on between all the people who are interested and supportive I don't think there is any danger of community gardens disappearing.[69]

The collectively produced knowledge is translated into power; it is a resource for the community in protecting its interests, and it transforms the position of gardeners in the local political structure. For years they were perceived simply as gardeners rather than activists; they belonged to social groups with no means and little political clout. Developing a better understanding of the urban power structure was the first necessary step towards developing their agency and become more significant social actors.

Realizing the position of gardens in the broader context of urban political-economy developed into a well-constructed critical understanding of the partnership between the City Administration and the private market. Gardeners learned a new set of concepts and practices that would enable them to fight for their gardens and develop an alternative vision for urban development. Although there were various efforts to create a broader collective before the battle against the Giuliani Administration's plan, it was mainly the imminent threat that facilitated the formation of a movement that shared a discourse. Sam from Harlem describes the initial formation of the movement and its discourse as an act of necessity since, for the most part, gardeners had to invent themselves as activists:

> People were coming from all over the city and it was a really critical time because many of us didn't know where to turn, what organization to turn to that will advocate for us. Because Green Thumb was a city organization they couldn't really advocate for us. So we had to find something that would speak to the issue, so we began to have a meeting at [the office of] Earth Celebration.[70]

Rene from the Bronx also recalls this transition from gardening to activism:

> In the Bronx we were gardening for years until we were threatened. We had to change our mindset and become activists. We had to learn how the city works. We had to look for gardens in the neighborhood since we realized that it is up to us—the community—that gardens would not be neglected. This was the reason that we needed neighborhood coalitions. If we worry only about our own garden we will lose. Our coalition, La Familia Verde, formed a farmers' market and established relations with schools and the church.[71]

---

69   Interview, December 5, 2005.
70   Interview, June 8, 2005.
71   Interview, November 2, 2006, South Bronx.

This transition from gardeners to activists is dependent on the production of counter-hegemonic representations of space that compete with the hegemonic one in order to then affect the material space.

A representative from Green Guerillas reflects on the process in which gardeners were developing into a stronger collective that disseminates knowledge and organizes action:

> The importance of the coalitions is related to political changes that happened in the city in the late 1990s when gardeners were isolated from each other. It wasn't helping them to preserve their gardens for the future. By working in coalitions, they still fight for their garden with their peers but also help each other out. It was the progression of time and New York City politics when it become more of a necessity for gardens to interact with each other more, on some levels, not all. They are still independent, different from each other, run different programs. Somehow, the coalitions that we helped to start in the different neighborhoods were a political act to straighten the voice of the gardens in a time that they were assaulted by City Hall.[72]

Ilya, an activist from the Bronx, talks about the realization of the position of the gardens in the city's power structure and how that led him and other gardeners to establish the More Gardens! organization:

> We started the whole movement at a time that gardens were well established and when gardeners started seeing what is coming down at the tunnel. Which was basically developers are very very hungry and city administration with politicians are all ready to really give the land that was fought for—the green community spaces—back for developers to be used for them.[73]

Ilya therefore suggests that part of the realization of gardener-activists was the workings of the city growth machine that acts against the gardens and the needs of local residents.

Gardeners and activists are well aware of the challenge that their actions and demands pose to market-oriented city politics. They see themselves, with pain mixed with pride, as the unwanted children of the city. Nevertheless, they developed new representations of space regarding urban political economy from their collective experiences of the gardens in the context of the city and the community. Eventually, they were also able to get some acceptance that in turn allows them to invest time and effort in better organizing themselves and in the development of their practice, discourse, and vision.

A dialectical process of action and knowledge, whereby one enhances the production of the other, was activated in the face of enclosing the material

---

72 Interview, September 29, 2004.
73 Interview, May 31, 2005, South Bronx.

128   *From the Ground Up*

space of the gardens. Gardeners had to learn the intricacies of the city in order
to outsmart it. They developed skills and critical knowledge and became aware
citizens who could read into the local political machine and counter it with claims
for social and procedural justice. Gardeners developed a broad understanding of
the dominant representations of space as well as alternative representations. They
also developed the mechanisms that keep the production of knowledge going in
order to translate it to power and protect the gardens in the future. They organized
themselves in neighborhood coalitions and in a citywide coalition of gardeners
(New York City Community Gardens Coalition), the main role of which was to
keep the dialogue on community gardens in motion; develop a strong collective
and educate it and the general public; and network with existing organizations that
act in the interests of gardens.

These new representations of space challenge some well-established notions
of (uneven) urban development and reverse the historically unjust distribution of
resources among urban neighborhoods. They propose instead an alternative set of
values based on the use-value of the space rather than its exchange-value—such
as the high value that gardens offer for the livelihood of people; their contribution
to social and cultural life; their role in improving neighborhoods and in creating
meaningful spaces for residents—while at the same time they discharge principles
of accumulation and capitalist practice values. This new knowledge is both a
collective resource for protecting the commons and a mechanism that defines,
shapes, and produces the commons.

Environmental Sustainability and Responsibility

Another theme entwined through the new representations of space concerns issues
of sustainability and responsibility for community gardens and the environment
at large. The well-developed notions that gardeners present in regard to these
issues are mutually reinforced by spatial practices (i.e. organic gardening, solar
energy, rainwater harvesting, education programs, and struggles over open space).
Environmental sustainability and environmental justice are directly read from the
engraving of community gardens in urban neighborhoods. Mr Abu, a gardener
from East New York, Brooklyn, captures insightfully the relation between gardens
and urban sustainability:

> It is vitally important to preserve all community gardens for the next generation
> because we cannot replace them. It is irreplaceable. We don't bring down a house
> to build a community garden so I think it is important that the leaders take a
> long-term look on the environment, quality of life, and on poverty. Yes, they are
> making money now on these houses, but 10 years from now it will create crime,
> it will create stress, it will create pollution with no open spaces. It is therefore
> vitally important that we preserve them and develop them as a community
> resource for people to use. You know there is a shortage of parks, shortage of
> community centers, shortage of …. This fills the void. We totally lack planning

towards making the city a beautiful space. You are walking down the street and it is house after house how about a beautiful lot that is a community garden?[74]

Ilya, an activist from the Bronx, presents a utopia of urban sustainability in which community gardens are seen as actually existing small-scale crystallizations of this vision, with their emphasis on green, community, and recycling practices:

We've done amazing on how to keep these precious jewels and we are now coming towards education and finding ways of creating more green space for schools and open spaces that are not necessarily buildable. There are still a lot of open spaces but they are grassy, parking lots, non-use rooftops. How can we move to the next level of opening community spaces that are there for parks? [...] The vision that we're pushing forward, I don't know if you saw the picture [a huge banner used by activists in garden events, rallies, and protests] where there is a building with a rooftop garden and a community garden on the side, and then water recycling inside the building, and solar panels on top, and then the next step is to take away the parts of the street to make it smaller for bicycles and trolleys and then making more green space in the streets, and sustainable buildings—just six floors, stepping it back. So we cut some part of them off and start bring dirt and sand and green and water filtration so the water can be recycled through plants and fish and then come down. So to make the whole planet, actually the city be green and sustainable so that a lot of the food and the water, and the waste, again composting toilets. So you don't need all these huge pipes to take the waste out and huge amount of pipes to bring water and huge amount of tracks and ships and trains to bring food. Some food brought from the Community Support Agriculture; you grow your own food and you have the water system, you have those beautiful rivers. All that could feed itself and stop destroying our plants and bombing Iraq and other countries and wasting so much resources and fighting over crazy reasons and start pushing in the way of making healing happen over the city and smelling fresh and butterflies and bees and animals coming back in.[75]

This vision that became the paean of gardener-activists entails certain environmental ethics and concerns. These were developed out of the strong and supportive emotional relations:

between local residents and the land they inhabit and come to care for. Such connections are a necessary condition for long-term environmental sustainability even if they are made to such humanly produced landscapes. These small plots connect us to our everyday environment in a *tangible*, rather than abstract, way. The point is not that we can therefore disregard the wilderness, but more that we

---

74 Interview, May 6, 2006, East New York, Brooklyn.
75 Interview, May 31, 2005.

must pay serious attention to the power of all environments *to draw us in as a partner worthy of protection.*[76]

There are some existing models which gardeners and activists draw from and adapt bits and pieces to their local urban setting and culture. There are established policies of protecting community gardens in other cities, and ideas of and experiments with urban agriculture and viable communities elsewhere. A representative of CENYC presents some of these inspiring models and their possible actualization in New York City:

> Internationally, urban agriculture is ramping, especially in Africa any unused piece of land on the side of the road, anywhere in a city in Africa has somebody growing stuff. Cuba too is an example for high percentage of fresh produce that people in some of these countries are able to eat all comes from stuff they grow locally themselves. That is a very hard thing to do in a city like New York because if you multiply the amount of space that you could possibly grow vegetables by the number of people and how much they eat, those two things are never equal. But there are other cities like Detroit that got acres of vacant land within the city. If they develop that land sustainably they could include urban farms that could grow a percentage of, it doesn't have to be everything, they are not going to grow the wheat for Detroit within the city limits but I bet they could grow every piece of lettuce that needs to be eaten in that city, every tomato that needs to be eaten, and most of the fresh vegetables because they have the vacant land. Philadelphia, another city that got a lot of vacant land slowly they are, I think Philadelphia is doing a better job than Detroit is doing with the land. But still they have a tremendous inventory of vacant land that if they thought about things in those terms of a sustainable city they could grow a large percentage of their produce right in the city limits. And those might be entrepreneurial farms within the city that could provide jobs and could provide income for the people who are doing the work. So that is the vision.[77]

But beyond developing feasible utopias of gardens and urban sustainability, a main mission of gardeners is to turn public attention—especially of the younger generation—to the significant role of the environment in their lives. Gardeners reach out to the community, adults, and children to share their experience of the gardens and the visions of sustainability they developed. They do so through various education programs (that were described at the beginning of this chapter)

---

76   Andrew Light wrote about the destruction of the Esperanza community garden in the Lower East Side (in February 2000). A court order to halt all further destruction of gardens until a compromise was reached was issued after Esperanza was destroyed. Light, A. 2004. "Elegy for a Garden." *Terrain.org: A Journal of the Built & Natural Environment* 15: 8 (emphasis mine).

77   Interview, December 5, 2005.

and through engaging activities in the gardens. One way to postulate the mission statement of:

> open space advocates and fresh food, people and organic gardening and all that kind of stuff is to have kids understand about nature, and about insects and about where food comes from, and actually physically experience it. Because just reading about it and watching a video about it or a television show or a movie have an effect on them but actually putting their hand in the dirt and finding a worm, picking an apple off the tree is an experience that I think is important people don't lose as the world become more urbanized. They [kids] come up with great things like—that is the crying tree—they heard about the weeping willow but they can also think about the fact that it is dripping down, like tears fall. They get a real feel for nature by being there.[78]

The gardens are a springboard for something bigger than the space itself. They provide urban children (and adults), especially the less affluent, a chance to really *experience* a diverse natural setting informally. The gardens offer a direct experience of nature, even if it is man-made; and, unlike urban parks that are too large and too homogeneous, they invite close observation and interaction.

Mike from the East Village suggests that communication of the kinds of knowledge and ideas of sustainability that have been developed by the gardens should not be limited to the younger generation but should also reach urban planners, developers, and urban scholars. The history of community gardens in his neighborhood envelops new ideas about urban development that can inform future practices:

> When I came here [to the neighborhood] I thought this is like Savanna in a way all these little pocket parks, Savanna, Georgia has these beautiful squares that really define the city. Everybody who goes to Savanna Georgia talks about those squares. And you know in my field and probably in your field it is always mentioned in the textbooks how important the squares were to the development of the city. And even though it [the gardens] happened after the development of the neighborhood, it is such an interesting way to redevelop a neighborhood, *infill parks instead of buildings.* So there is got to be a way to grasp onto that and use that as a way to market the neighborhood but not in the sense that it is just about tourism in a sense that it builds the sustainability of these gardens is the really important part of the neighborhood. [...] 'cause it is really about sustainability and this neighborhood is a textbook example of that not just because we have so many gardens but because the gardens just seem such [an] innate, natural part of it. And I think that has to do a lot with the geology of this

---

78   Ibid.

site and the fact that it was never really meant to be built on and never really meant to be built on in this density.[79]

Community gardens are captured here as providing a valuable roadmap for sustainable practices that cities of the future will need to have.

Alongside such all-encompassing visions are also smaller, more concrete ideas and discoveries on the issue of health and environmental sustainability, small discoveries that make the prospects of the bigger vision more feasible— recycling and composting, rainwater harvesting, permaculture growing, etc. Ilya, for example, discovered a mushroom in the garden that can cleanse the soil of the toxic residue of car oil:

> I find mushrooms that we planted were they spill tons of car oil running into the garden killing one of our trees at the front. So we found out those people from California that use this fungus oyster mushroom which is embedded into wood chips in the garden, and eats the oil, which on the molecular level turns into food. So it is not just bring[ing] the toxic out and then you have to throw the mushroom away; it actually breaks down the oil and eats it as edible food. So that is a huge. [We need] more plants that can break down all these toxics that were created and turn them into food. And that will be, I think a huge next step of healing and also make money out of the waste that was created in the past. So that is the next step, finding ways of healing and creating things.[80]

Mike from Manhattan also talks about healing, the healing of a community and the individuals in it rather than the healing of the soil:

> I definitely think that having this green space is really valuable to urban dwellers and neighborhoods as dense as this. Just being able to look at a tree reduces stress and is therapeutic. So that is the very basic thing and that is what got me involved. The thought of losing the view that I had of those willow trees, it scared the hell out of me. It was the only thing I really liked having down here because it really was a bad neighborhood. And it was even worse before I got here. So there is that and then all the contribution of just the trees and the amount of oxygen that a tree as big as this willow or linden can produce, and the amount of carbon it can take in. All of those things, the ecology, the wildlife habitat, and other, more technical ecological things that we know about in an abstract way that contribute to the health of the community.[81]

The question of sustainability as it pertains to community gardens in New York City is subject to controversy, which is explicitly apparent in the different agendas

---

79   Interview, October 5, 2006, East Village, Manhattan.
80   Interview, May 31, 2005, South Bronx.
81   Interview, October 5, 2005, East Village, Manhattan.

of the community gardens support organizations. There are different views on what is the better way to sustain the gardens: securing the land or securing the community of gardeners. But gardeners themselves present a much more elusive and tacit approach to the issue of community garden sustainability. They have several ideas about what needs to be done in order to keep the gardens going and to even grow in numbers. They have strategies and mechanisms in place to protect the gardens from development or from becoming inactive. Some gardeners are more oriented towards one side of the controversy, some towards the other. Some are trying to combine the two poles into one holistic approach. Regardless of their orientation, garden coalitions and supporting organizations are working to establish the gardens as an integral and inseparable part of the city, entrenched in the collective perception and practices, laws, and local politics. This effort towards the institutionalization of community gardens is directed at the future existence of community gardens. But the sustainability of gardens begins from the mere production of space in a certain way: by (re)creating the commons—the actual space in which a new collective wisdom can develop and flourish.

Construction of Class

There is a shared understanding within the collective of gardeners of the specific social class that the gardens serve and who fights to protect them as an asset. The notion of social class here distinguishes between the haves and have-nots, between the powerful and powerless, between the rights-full and rights-less, and between the owners of land and its non-owner users (and in the case of the gardens also the producers). Neither Marxian nor Weberian definitions of social class are flexible enough to encompass all these meanings. An alternative conceptualization defines class not by the criterion of ownership—who owns the means of production—but by groups' position in a specific historical configuration of power and powerlessness. According to this broader understanding, class is always in a process of formation and articulation. As such, class formation and articulation are the means of the struggle of the powerless as well as its end.[82]

The construction of class may be most explicit in claims for material space, or for its more egalitarian distribution. Community gardeners live in neighborhoods that are disproportionately and unjustly deprived of green open space and that do not satisfy the need for recreational environment; neighborhoods with an overwhelming majority of renters rather than homeowners (see Map 1.5, p. 30); and neighborhoods with a high concentration of lower-income households (see Map 1.3, p. 28). Gardeners are using claims of unjust distribution (of space and resources) as part of their struggle. They also refer to the discourse of unjust distribution of power, with demands to be heard by decision makers and to have more control over issues pertaining to their gardens, their neighborhoods, and their lives.

---

82 Aronowitz, S. 2003. *How Class Works: Power and Social Movement*. New Haven: Yale University Press.

Claudia from Harlem talks about the imbalance of power that her fellow
gardeners, most of whom are immigrants from Mexico and Puerto Rico, experience:
"It is hard for somebody in that position who not only doesn't have a lot of education
but now finds himself an immigrant in a foreign country." Though she is white and
American-born, she can observe very intimately what it means for others from less
privileged backgrounds with fewer opportunities. Moreover, she is also part of the
have-not group herself. In her daily life she lacks access and opportunities to places
of recreation and social gathering other than the garden. She suggests:

> When you think about it, there are not many places in the city where people get
> together that are free. We are getting together here to eat a meal at the restaurant
> but we are paying for it. Having a place to get together with neighbors and
> friends and just relax and talk all day, it is either in New York City out on the
> sidewalk or in a park or in a garden. Unless it is a social club and there is a
> particular group of people who belong to social clubs.[83]

The tension between the owners of land and the users/producers of land becomes
noticeable in neighborhoods which are targeted for gentrification in strong feelings
of resentment towards the newcomers: "all these people with nice brand new shiny
cars and brand new houses for a few hundred thousand a unit." Gardeners witness
these social and spatial changes in their neighborhood and generally feel that they
"are not getting their fair share and the little bit of *land that they are trying to grow
up also being yanked away*." But the real anger and blame is directed not at those
well-off new residents, not at the local manifestation of the problem, but rather
through a broader understanding of the city's "vision [that is] really replacing
the people who have been very hard working, working class and barely existing,
through this huge gentrification."[84]

Because they belong to poor, usually physically destroyed and generally
underserved neighborhoods, they are the target of either gentrification or over-
densification with "crap housing" and various treatment institutions (i.e. rehab
centers, senior centers, etc.) that bring more poor, underprivileged people to the
neighborhood. The idea of class is consolidated through the space of the garden as a
resource for the traditionally resource-less and rights-less. It is because the space of
gardens is perceived by gardeners in class terms—as a resource—and if this resource
is being threatened by the powers that be it is something worthy of fighting for.

Various aspects of social class are articulated by gardeners. The human
diversity that is present in the gardens has already been mentioned but should be
reemphasized with regard to social class. Mike from the East Village talks about
this diversity in his garden: "We get people from that strata, from Christadora
[building] and stuff like that, people with money. And then we get people like T—.
I don't know if you know her, she lives in section 8 housing on disability and she

83   Interview, November 8, 2005.
84   Interview, May 31, 2005, South Bronx (emphasis mine).

is totally crazy."[85] Income levels among gardeners are much more diverse today than in the 1970s and 1980s, mainly as a result of neighborhood gentrification. But for gardeners who are better off, participating in the garden allows a certain social gaze on the lives of others in a way that is usually unavailable elsewhere. In my own experience as participant-observer—being a gardener and working with the New York City Community Gardens Coalition—I developed personal relationships with people whose lives are vastly different than mine and who I probably would not encounter otherwise.

Among the poorer gardeners, those whose housing is not secured by some kind of government subsidy, there is a mostly silenced faction that was already displaced from the neighborhoods that were gentrified. Such is the story of Bell, whom I met (regularly) in a garden in the East Village. Bell recounts:

> [The] high rent drove us out. I was here 17 years before I was let out [...] we went from sublet to sublet, but with children it is hard to make so many changes, so we found a place in Queens. But we come here, this is really our neighborhood, because our neighborhood in Queen is very ... there is no public place except for the park.[86]

The story is even more cruel and tragic considering the assertion made by Emily from the East Village:

> So many people use the gardens and in fact I think that is one of the reasons that market value of apartments and stores went up in this neighborhood. So much was because of the gardens, because a lot of people will pay a little extra to live near a garden.[87]

This assumption is bound by a study on the effect of community gardens on property value in the Bronx. It showed that in all the neighborhoods, the presence of gardens significantly increases the value of the surrounding property, both residential and commercial (within a 1,000-foot radius). In some cases the increase was up to 9.5 percent within a few years of establishing the garden.[88] This puts community gardeners in a paradoxical position of having a role in their own displacement.

There is another inner conflict that is generated by the gentrification of neighborhoods. The incoming residents increase the social diversity of community gardens but also introduce different needs of different social groups. These better-off residents seek to realize different values of the gardens. For example, they wish to emphasize recreation and education over food production. In addition, since

---

85   Interview, October 5, 2006, East Village, Manhattan.
86   Interview, August 19, 2002, East Village, Manhattan.
87   Interview, April 24, 2002, East Village, Manhattan.
88   Voicu, I. and V. Been. 2008. "The Effect of Community Gardens on Neighboring Property Values." *Real Estate Economics* 36(2): 241–83.

they are more likely to be homeowners than renters they might put more emphasis on the garden as leverage to raise the value of their property. That, in turn, might encourage them to care more about the look of the garden (possibly even leaving the design to a professional) than about the participatory process of its design. A diverse class composition, then, also leads to diverse and even conflicting needs.

Class formation of urban gardeners is an important part of transferring power to the powerless, a goal of both gardeners and organizations. One prominent example is the effort of TPL to make the gardeners the owners of their gardens and endow them with a position of power that they usually do not occupy: as the dominant members of the board of directors of the land trusts. Instead of filling the board of directors with "people from planning departments, real estate companies, and foundations" and reifying the constellation of power, it would bring the traditionally powerless to the board (from among the gardeners themselves)—an act against convention in regard to power entitlement.

We have seen that one of the representations of the garden, which is produced by the gardeners, is as a space that defines and is defined by social class: a space with which a clearer distinction of the powerless from the powerful is constructed, and a space to which visions of alternative construction of class are attached. The knowledge that is being produced and shared in community gardens amounts to a discourse that informs the collective of gardeners and is reinforced by them. This collective wisdom—which is in dialogue with issues of class construction, uneven distribution, urban political economy and development, and questions of sustainability—symbolically unites gardeners across the city (despite the fact that its production happens, for the most part, very locally). Gardeners at the forefront of the movement are aware of the power of this knowledge to galvanize the gardeners and bring them together. Therefore, they see themselves as the educators of the collective of gardeners and of the public at large. For the gardeners who follow the leaders of the movement this knowledge is an opening for connection and belonging to a bigger and (for a change) more powerful group.

More importantly the collectively produced knowledge shapes the social understanding of gardeners and offers them a more powerful stance with regard to social issues. There are values that these gardeners, usually marginalized and silenced, are getting personally and collectively from being part of a garden in a way that keeps them from submitting to the hegemonic knowledge. This new knowledge is their way of using their own resources for themselves, producing a discourse that supports their own sustainability (as gardeners, as urbanites, etc.) but also producing something valuable to others: producing and sharing knowledge that is concurrently counter-hegemonic and constructive. This knowledge challenges existing representations and frameworks of urban development, sustainability, and social inequalities. But the alternative that they propose can serve and inform city planners and community organizers and other professionals who tend to approach neighborhoods and communities top-down and impose upon them representations that are derived from the dominant framework.

Conclusion: Connecting Three Moments of Space

A community garden is not a single-person project: it necessitates a collective act of people working together to change their life conditions, to create something of their own, to inspire and be inspired. It is a collective act of people who have recognized the deficiencies and problems in their physical and social environment and are able to find the resourcefulness (in mind and action) to alter these problematic conditions. These practices can be seen as *reworking* practices which entail and proceed from recognition of problematic life conditions. This recognition generates actions in order to "recalibrate power relations and/or redistribute resources" and to "undermine [...] inequalities on the very grounds on which they are cast." Reworking practices, then, are actions that respond to a problem on the same level and scale that it occurs. It is a local response to a locally identified problem, rather than a broader response to global social problems of inequality and oppression. Nevertheless, reworking not only affords more livable, workable lives for those engaged in the practice but also has some broader implications. First, through these acts people reinstitute or retool "themselves as political subjects and social actors."[89] Second, these practices further facilitate the "recognition of what was privileged by hegemonic social forces,"[90] and therefore generate a dialectical process of action and recognition. In the process of acquiring the skills to change their own life conditions, actors are becoming familiar, in a more critical way, with the world around, how it operates, and what constitutes power in it.

The reaction of residents to life conditions in New York City in the early 1970s resulted in the inauguration of the first community gardens. Their actions to rework their living conditions—their transformative practices—both constituted them as social actors interacting with and modifying their living environments and allowed the emergence of a discourse that recognizes the hegemonic forces at play. Residents faced devastated environments and decided to do something about it. They retooled themselves with the skills and know-how to transform and heal the conditions that victimized them. Their resourcefulness and direct action are derived from a realization of their own position within the city's political economy. They recognized that their neighborhoods were generally deprived of capital investment, municipal resources, and support. And they distinguished the powerful from the powerless—the haves from the have-nots—and realized that it was up to them to find solutions.

Reworking practices are not limited to the inauguration phase of community gardens but continue to hold sway. Residents taught themselves and learned from each other how to become gardeners and produce vegetables and fruits in an urban environment. These skills are particularly important where fresh

---

89   Katz, C. 2004. *Growing Up Global: Economic Restructuring and Children's Everyday Lives*. Minneapolis: University of Minnesota Press, p. 247. Reworking is the second "R" in Katz's triad construction of resistance (resilience, reworking, resistance).

90   Ibid., p. 249.

produce is scarce: poor, inner-city neighborhoods. They also learned the practice of managing a community garden, a practice that entails negotiating between diverse needs and personalities in a collectively owned space, and sustaining the garden in the face of changing levels of participation, resources, and city and community support. The various understandings and visions in regard to urbanity, community, natural environment, and social inequalities—along with the mechanisms to share and disperse the knowledge—were developing gradually out of the transformative acts of participants. All these efforts to alter the life condition of marginalized communities in New York City are directed towards finding a solution on the same scale as the problem. The formation of the gardeners' collective and its production of the commons to provide lacking but needed resources do not redress and challenge the broader social injustices that brought about these conditions. Nevertheless, in the case of community gardens there is still a potential for these reworking practices to develop and expand their scale to redress social and spatial processes that determine life conditions in local places.

In an attempt to revisit the notion of the commons not just as a utopia but as an actually existing space amidst the neoliberal city, I reviewed the commons to be reproduced by three interconnected elements: material space, knowledge, and meaning. The material space of the commons is produced, maintained, and protected collectively by its users. The value exerted from space compensates for and supplements needed but unavailable resources. The knowledge pertains both to the practical knowledge that enables the ongoing production of the commons and to the discursive framework that defines the commons. The lived space of the commons, exemplified here through the reinstitution and celebration of various cultures, also encompasses an alternative aesthetic experience that challenges aesthetics norms; an alternative social experience that challenges the prevailing alienation of people from their physical and social environments; and an alternative psychological experience that thrives on an enhanced sense of control and belonging. The existence and persistence of the commons depend on these three interrelated elements; each element constitutes and propels the others, and together they enable urbanites to constitute an alternative urban experience.

The side-by-side positioning of actually existing commons and "actually existing neoliberalism"[91] constitutes the locale as a contested arena of opposites, ambiguities, and as a paradigmatic site for the examination of struggles over space and the spatially embedded potentialities for social change. We can understand the practices of producing the commons on two different levels. First, they can be understood as the practices of "subaltern class actors [that] struggle over the appropriation of material and symbolic goods."[92] That is, a collective action aimed

---

91   Benner, N. and N. Theodor. 2002. "Cities and the Geographies of 'Actually Existing Neoliberalism.'" *Antipode* 34(3): 349–79.

92   Aronowitz 2003, p. 51.

at receiving a greater share of goods without challenging the social mechanisms and institutions that produce inequality. Goods and rights such as open space, clean neighborhoods, healthy food, and ownership were unevenly distributed and deprived of the collective of gardeners. The commons, then, is a mechanism for redistribution through which underprivileged residents compensate themselves for uneven urban development.

The second level at which to understand the production of the commons is as a collective action that challenges the hegemonic social order and follows instead an alternative logic of justice.[93] By producing the space of the gardens, gardeners present a defiant and provocative alternative to the dominant social space—an alternative that redresses the right to public space not only in its concrete sense but also in the "the right to the city" sense. It is an alternative to the logic of organization and planning of space, to the distribution of control over it, and to its meaning and experience. By introducing alternative practices and values to capitalism, the commons are de-enclosed and the dominant mode of production is challenged.[94]

Alternatives to capitalism (with its narrowing down of space to property) such as the commons are constantly under threat of being enclosed and becoming a generative force of capitalist reproduction. It is possible to strip community gardens of their critical potential and make them a force of reproduction of the dominant system rather than a force of transformation.[95] Nevertheless, as we are now standing at the threshold of a new era, alternative modalities of social reproduction that take after the model of the commons are increasingly being developed.[96]

Community gardens as actually existing commons offer a glimpse of the kind of social relations and spatial practices and values that can bring the commons back to everyday urban life. They facilitate a cooperating and participating community, gathered around non-commodified activities, collectively producing space according to their needs and visions.

Community gardens, then, are sites for re-visioning the urban environment as the commons. Because "simultaneously present in any landscape are multiple enunciations of distinct forms of space—and these may be reconnected to the process of re-visioning and remembering the spatialities of counter-hegemonic

---

93   Ibid.

94   De Angelis, M. 2007 *The Beginning of History: Value Struggles and Global Capital*. London: Pluto Press.

95   For discussions on community top-down gardens that are not necessarily serving the common wealth, see Eizenberg, E. and T. Fenster (forthcoming). *Whose Power, Whose Autonomy? The Contested, Controlled and Neoliberalized Community Gardens*; Pudup, M.B. 2008. "It Takes a Garden: Cultivating Citizen-Subjects in Organized Garden Projects." *Geoforum* 39: 1228–40; Quastel, N. 2009. "Political Ecology of Gentrification." *Urban Geography* 30(7): 694–725; Rosol, M. 2011. "Community Gardens as Neoliberal Strategy? Green Space Production in Berlin." *Antipode* 44(1): 239–57.

96   De Angelis 2007; Hardt and Negri 2009.

cultural practices,"[97] the entrenchment of an alternative vision in actually existing spaces (in the material space) offers the potential for transformation of some fundamental aspects of everyday life and social organization. Actually existing commons then should not be seen as a "return" of some noble but possibly archaic ideal, but as a springboard for critiquing contemporary social relations and as the production of a new spatiality, initiating the transformation of some fundamental aspects of everyday life, social practices and organization, and thinking.

---

97   Keith, M. and S. Pile. 1993. "Introduction, Part 1: The Politics of Place." In M. Keith and S. Pile (eds). *Place and the Politics of Identity*. New York: Routledge, pp. 1–21, p. 6.

PART III
Reaped Politicization

Why should the ultimate victory of a trend be taken as a proof of the ineffectiveness of the efforts to slow down its progress? And why should the purpose of these measures not be seen precisely in that which they achieved, i.e. in the slowing down of the rate of change? That which is ineffectual in stopping a line of development altogether is not, on that account, altogether ineffectual. The rate of change is often of no less importance than the direction of the change itself; but while the latter frequently does not depend upon our volition, it is the rate at which we allow change to take place which well may depend upon us.[1]

Polanyi's questions and conclusion about efforts to present an alternative to the dominant trend help recontextualize the actions of community gardeners as efforts to slow down the continuous neoliberalization of urban space. In concrete spaces in the city they brought the progress of the neoliberal plan to a halt not only in the actual spaces of the gardens where plans and policies were not executed, but also in the counter-discourse and counter-actions to the neoliberalizing *zeitgeist* that flourishes in those spaces.

There are two different approaches to conceptualizing efforts to present alternatives that can be located on either side of Polanyi's assertion. In the first approach, counter-hegemonic actions are perceived as mere individualistic acts of resilience. These acts release some of the tension caused by oppression, but are not presenting a real political alternative and will not amount to a force that can transform the oppressive nature of a hegemony.[2]

The second approach to the construction of an alternative suggests that counter-hegemonic forces that act on space are aimed at inducing some actual modifications. Therefore, by transforming space the social relations that produced the space and are reproduced by it can be altered. In other words, social change (rather than merely individual relief or the relapse of dominant progression) can come about through forces reclaiming space and changing the conditions of its production. Space is the locus from which the promise of liberation from repression and exploitation can spring. Furthermore, the potential for change lies

---

1 Polanyi, K. 2001. *The Great Transformation: The Political and Economic Origins of Our Time*. Boston, MA: Beacon Press, p. 39.

2 For example De Certeau, M. 1988. *The Practice of Everyday Life*. Berkeley: University of California Press.

in differential space—where differences and the 'right to difference' are being celebrated—that allows for alternative ordering of the social and alternative life experience. The acts of production of differential space can amount, according to Lefebvre, to a comprehensive transformation of society.[3]

Looking for indications of transformation, the following examines the institution of community gardens in New York City as a political body that carries out efforts of change. We have already discussed the formation of the collective of community gardeners and some of its products (i.e. the commons) and we shall now see how this collective has been, and is being, politicized through the process of institutionalization. The institution of community gardens configures ways—practical and discursive—to confront the dominant trend that works against it. It is constituted as another platform for the processes of political development of residents for regenerating the transformative capacity of the movement.

---

3   Lefebvre, H. 2003. *The Urban Revolution*. Minneapolis: University of Minnesota Press.

Chapter 6
Rooting Politics: The Institution of Community Gardens

In what ways can the phenomenon of community gardens in New York City be viewed as a counter-force within the urban environment, striving to resist and transform the prevailing social and spatial environment? The process referred to here as the *institutionalization* of community gardens represents the struggle of the gardens to obtain more power and influence in the city's affairs and to extend their impact on spatial and social issues beyond the premises of the gardens. More specifically, in the process of institutionalization, everyday spatial practices are translated into political action through the formation of institutions.

The political practice and discourse of community gardeners is bound to be anchored in a material space and in everyday practices, an anchoring crucial to the construction of a new powerful organization of the locale. Other examples of urban organizations that strive to maintain locale autonomy—such as workers' cooperatives, limited equity cooperative housings, and banking cooperatives— were found to have two major flaws. First, they tend to see themselves as "opting out," as alternatives rather than oppositional. They employ an additional set of practices, alongside mainstream ones, which dooms them to remain politically marginal. Second and related, they gradually "mainstream" themselves due to their dependency on government and nongovernment grants. Therefore, it is not enough to operate based on alternative principles of ownership: reflexivity and consciousness of struggle are also needed in order to engender a broader, more radical transformation in people's understanding of the world. To foster real social change such organizations need to develop oppositional politics, become independent of outside capital resources, and become part of a political movement.[1]

The institution of community gardens is interesting for its structure and its practice. Its components were discussed in previous chapters (mainly Chapter 4) but are being recapped and elaborated upon here. The practice of the institution involves maintaining the right to participate in the urban sphere, to affect it, and to protect the spaces that are an important part of everyday life. These practices—strategies that guide the institution—begin in the realm of the garden and stretch to encompass actions that are broader in scale (neighborhood, city,

---

1   DeFilippis J. 2004. *Unmaking Goliath: Community Control in the Face of Global Capital*. New York: Routledge. DeFilippis is interested in the potential of forms of local ownership that are anchored in space to re-fix capital to space and create a powerful locale vis-à-vis global forces that tend to dispossess local capital.

state) and in essence (targeting issues that are not limited to the needs of gardens themselves). The institution's strategy of struggle, it will be shown, is not fixed but is changing and developing according to the historical context. When gardens face imminent danger the strategy is more concretely directed towards protecting material spaces. In periods of more stability the strategy is more oriented towards further institutionalization of community gardens as an effective power within the urban power structure. Finally, this chapter discusses the tension that is present in the development of the institution of community gardens in the city: a tension between the grassroots character of the movement and its decentralized and open structure that promises independence from affiliation's claws on the one hand, and the institutionalization of the movement that promises more stability and political clout but involves risks of mainstreaming on the other.

The Pillars of the Institution of Community Gardens

The institution of community gardens is comprised mainly of the different units of the collective. These include the basic horizontal network of community gardens (standing today at about 650). Each represents individually a collective form of ownership, and together they cover about 90 acres of land. Attached to the basic horizontal network is a more complex vertical network.[2] Gardens can be part of a *neighborhood coalition*. Neighborhood coalitions are among the most important vehicles of establishing the garden as an indispensible neighborhood institution. The coalitions provide a supportive network for gardens within the same locale, where individual gardens collaborate in order to act more efficiently. As the leader of the Harlem United Gardens Coalition explains, "Twelve gardens getting knowledge, funds, and political support are stronger than one garden."[3] The coalitions are important also as "gatekeepers" for the gardens in the neighborhood. They centralize information and orchestrate action in regard to developing threats and neglected gardens in the area. Finally, they are an important means to connect the gardens to other neighborhood institutions, public and private, through ongoing relations with them. Neighborhood coalitions collaborate with schools (hosting classes on gardening and environmental education, and many schools maintain a garden plot); soup kitchens (providing fresh produce or helping to distribute food); churches (hosting religious events and celebrations); block associations and

---

2   This network can be seen as part of a larger, nationwide network—the American Community Gardens Association. However, the discussion in this chapter is limited to garden institutions in New York City and maintains the institutional relations beyond the premises of the city, in this case, as less relevant to the understanding of urban space, power, and politics.

3   Interview, November 22, 2004, Harlem, Manhattan.

community boards; farmers' markets, banks, hardware stores, and more.[4] These collaborations enable the integration of the gardens into the "operative network" of the community and strengthen their position as important neighborhood institutions.

Forming neighborhood coalitions and "branching out" to other neighborhood institutions is a means to achieve greater local autonomy and appropriate "the structures and institutions that are the focal points of those [mundane] experiences and practices."[5] Through these acts of networking and branching out individual gardens and neighborhood coalitions generate an impact on the neighborhood that exceeds the actual space of the gardens and enables community residents to participate in an assemblage of institutions that are part of their daily life and that affect them. Such acts of civic engagement are a means by which social capital is engendered and in turn reinforces the power and capacity of the locale.[6]

Another pillar of the institution of community gardens is the remobilized New York City Community Gardens Coalition (NYCCGC). This nonprofit organization was founded in 1996 and incorporated in 1998 predominantly to respond to threats to community gardens around the city. The establishment of this citywide umbrella organization was triggered by the gardeners' perception that there was no other organization that could assume the role of advocacy for community gardens. Dealing with threats from the City Administration and the private market, other supporting organizations could not openly advocate for community gardens and risk their own private and municipal support.[7]

New York City Community Gardens Coalition used the office space of another organization—Earth Celebration—in the Lower East Side of Manhattan and began to orchestrate rallies, demonstrations, and outreach events targeting the general population and politicians advocating saving the gardens. Sam, the first president of the coalition in 1996 and a gardener from Harlem, describes the first steps of mobilization:

> We just began to build [ourselves] and as the struggle increased around the
> gardens we began to be more active in terms of going to rallies, going to the

---

4   In a self-report survey from 2006 and 2007, 31 percent (in 2006) of the gardens reported collaborating with local schools (37 percent in 2007); 13 percent (2006) and 11 percent (2007) with soup kitchens and food pantries; 50 percent (2006) and 40 percent (2007) maintained contact with the community board, of which 22 percent had gardeners enrolled on the community board; more than a quarter of the gardens had representatives in block associations.

5   DeFilippis (2004, p. 10).

6   Saegert, S. 2006. "Building Civic Capacity in Urban Neighborhoods: An Empirically Grounded Anatomy." *Journal of Urban Affairs* 28(3): 275–94.

7   Green Guerillas was an exception. It helped the coalition submit the lawsuit against the administration that generated the court order to stop the destruction of gardens until a proper public review process was conducted. Green Thumb's staff clandestinely informed gardeners about upcoming threats and helped them organize their coalitions.

elected official and demonstrating in front of their offices, going to city council hearings, land use and parks and recreation [meetings]; it was mainly land use at the time. We were trying to keep focus on the different sites that were coming up for development and were being developed. We were trying to keep a focus on what their status was. So we concentrated on the gardens and we eventually put together a bulldozer hotline where, if a garden was threatened, they could call the bulldozer hotline and people would come out and sit at the site or demonstrate at the site, whatever was needed. So we really began to get involved in terms of working, struggling, and saving gardens.[8]

In 1999, the threat from the Giuliani Administration became more imminent and many more gardens were slated to be auctioned. At that time Françoise Cachelin, a founding member of Creative Little Garden,[9] was part of the coalition—and her apartment even served as its headquarters. She founded the "bulldozer hotline" for threatened gardens to call and recruit people for demonstrations and garden sits-in. For many people, this over 80-year-old lady with endless energy and the ability to bring people together, strategize, and carry out the missions was the leading light of the struggle.

Sofia from Creative Little Garden befriended Françoise. Though she was not one of the leading activists in the coalition, Françoise made sure to keep her involved. Her recollection of those days provides a sense of the spirit that characterized the initial moment of gardeners' organization:

Because she was so overwhelmed, it was during Giuliani, when a lot of the gardens were being lost to developers, and some of them were so beautiful it was literally a crime to see them go. They had full-grown trees; they had incredible landscape, amazing. So she basically created the New York City Community Gardens Coalition, with a bunch of people. So she basically asked me to come; they planned a very big event called "standing our ground" and she asked me if I want to help out. That was 1997 and in the neighborhood there were a lot of gardens right here that were being destroyed. She called me and said, "We need people to go to city hall today, there is going to be a hearing the more bodies […] or there are bulldozers coming." So I would go there. That is how she got me involved. And she wouldn't take no for an answer; she would call and say, "If you don't come the gardens will be gone" and she had that kind of force she would convince anybody. She was a little woman, 5' 2"; she was bent but with just so much enthusiasm and drive to get things done. The phone number for the coalition was her apartment so people called her house. She was retired so this was her job. I admired that. She is really the one who got me involved.[10]

---

8   Interview, June 8, 2005, Harlem, Manhattan.
9   Located on 6th Street between A and B Avenues, Manhattan.
10   Interview, October 7, 2005, East Village, Manhattan.

Françoise Cachelin died in 2003, and with the victorious results of the struggle many organizations within the community garden world, including the NYCCGC, receded and lay dormant for a while. Those gardens that were preserved were content with their victory and gardens that were slated for development mourned their loss with a sense of betrayal by fellow gardeners of preserved gardens. The victorious settlement divided the gardens, and sporadic battles were fought on a case-by-case basis or at neighborhood level at most.[11] But the institution was an established fact, formulated with ideas and practices. It had its own bylaws that defined its mission and conduct as well as five years (until 2003) of monthly meetings of members and of the board of directors, documented in formal minutes alongside all the actions and events that they orchestrated throughout and since the struggle. Sam from Harlem explains:

> We need something for the long run. We need an organization like the [citywide] coalition to pay attention to that and to educate the public to the importance of that. And not just being gardeners, we are also guardians of the land so it is really important to pay attention and stay in focus. [...] We need to study and keep ourselves abreast with the issue and watch the land use [plans] and we really need to maintain our gardens. [...] Basically it is reaching out, trying to tap into existing organizations; there are lots of organizations that claim to be acting in the interest of gardens and open space and green space. So we are trying to plug into those different organizations, and trying to pull all that information, all these resources, pulling them together, not so much keep them on. But just to try to keep things accessible to garden groups.[12]

Gardeners realized the need to develop a broad understanding of the dominant representation of space, and needed to keep the production of alternative knowledge in motion in order to translate it into power and protect the gardens.

However, though there is a preservation settlement in place, issues of gardens' future survival are still arising. At the same time the decline in activity and connectedness of gardens took its toll. In 2008 the reservoir of gardens had shrunk from over 100 gardens designated for development in 2002 to about only 50 left undeveloped. In addition, despite the required public review process—the Uniform Land Use Review Procedure (ULURP)—before a garden site is earmarked for development, most of these gardens were snatched away without due process and with none of the government agencies, neither Green Thumb nor the Attorney General's office, enforcing it. At the same time, the city was not issuing leases for new gardens to emerge.

---

11   Such was the case of More Gardens! Organization's resisting for two years the plan to build on the site of three gardens in the Melrose neighborhood, South Bronx. Eventually, in November 2004, they lost the battle and with it some 11,000 sq. ft of garden space.

12   Interview, June 8, 2005, Harlem, Manhattan.

Realizing the toll of this decline, the first seeds of remobilization were planted by several neighborhood coalition leaders who initiated, planned, and in 2005 carried out a speak-out forum for gardeners citywide. The purpose of the forum was the construction of a stronger institution of gardens even during relatively secure times. The organizers of the forum wished to open a dialogue between gardeners and the various related organizations on the ongoing issues that gardeners were facing, and on the future of gardens in the city. The goal, as they defined it, was "to provide an environment for community gardeners to collectively establish needs and priorities. These needs will be taken to the city by a smaller number of community gardener representatives."[13] In order to get gardeners involved in the process of defining the agenda for the event, a survey was created and disseminated before the forum. On a stormy Saturday of April 2, 2005 about 100 gardeners from all over the city gathered in mid-town Manhattan to hear answers from representatives of related organizations[14] and to brainstorm solutions for the major challenges facing gardeners.[15]

The forum was hailed as a success by its organizers. The guest politicians could only be impressed by the seriousness of the event and the high rate of attendance. The success of the event also showed the organizers that the gardens movement had not lost its strength.[16] The end product of this effort was a newsletter reporting the content of the forum that was disseminated to all the gardens.[17] Overall, the forum served as a platform for networking and collaboration between gardeners citywide and as an opening up of the opportunity for the gardens "to become institutionalized."

The board of directors of the citywide community gardens coalition that had been inactive for almost three years (since 2003) realized the danger of losing the reins and announced its renewed activity at the forum. The first NYCCGC membership meeting, in May 2005, followed the forum by a month. The meeting called for members from all over the city to reunite their efforts and form a proactive group that could help protect the remaining endangered gardens and envision preparing a sustainable future for all gardens. In addition, the 2010 end-date of the settlement with the city was approaching and required solutions. The president of the coalition at the time, Mr Austin, a gardener from the Bronx, laid out the vision for the coalition in that first meeting as follows:

---

13   March 10, 2005—Forum Planning Committee, meeting notes.

14   At the First Annual Gardeners' Forum representatives from the following organizations participated: NY State Attorney General Office, NYC Parks and Recreation Department, Green Thumb, the Department of Housing Preservation and Development (HPD), Neighborhood Open Space Coalition (NOSC), the Trust for Public Land (TPL), and the New York Restoration Project (NYRP).

15   The work groups were developed based on a preliminary survey of problems and included: insurance, membership, endangered gardens, infrastructure and materials, and fundraising.

16   Internal communication among the organizers of the forum.

17   "What's Bugging You?"—newsletter on the community gardens forum, April 2, 2005.

We want to build our muscles, the more we have the stronger we become. In the Bronx we were able to save three out of ten gardens; that might sound little but it is a lot. The gardens will be up for review in five years, we should not be a reactionary group but an action group. Act now to prevent from failing in the future. We want to strengthen ourselves. First, to show numbers so they will listen to us.[18]

The coalition stressed the danger of resting on the laurels of the settlement, a tendency among gardeners and activists. As another board member suggested in the same meeting: "gardeners are again becoming complacent but they cannot relax. We should continue our advocacy, being community organizers and leaders. We should be always on the pulse of city policy, planning in order to protect our community."[19]

The coalition's overarching goal "is to advocate for community gardens preservation in New York City." Its set of missions is defined as follows:

1. Promote the preservation and creation of community gardens and community-developed open space in the five boroughs of New York City.
2. Educate the public about the value of gardens and the benefits they confer on New York City residents.
3. Serve as an effective resource for providing information and technical support to community gardeners.
4. Raise the profile of community gardening among elected and appointed officials, the media, the environmental community, and the general public.
5. Foster networking and communication among the NYC garden communities.[20]

NYCCGC meets every month, with varying levels of participation. The meetings usually involve a discussion and an update on the various issues gardens are dealing with at the time. Several action committees operate on behalf of the coalition:

- the forum committee—orchestrating the annual gardeners' forums since 2005;
- legislation committee—working on long-term sustainable solutions for the gardens, and in particularly preparing a response to the end of the gardens' settlement in 2010;
- fundraising committee;
- the more recent advisory committee to Green Thumb (since November 2007).[21]

---

18 NYCCGC meeting, May 18, 2005.
19 Ibid.
20 http://www.nyccgc.org (accessed May 7, 2007).
21 The actions and visions of the neighborhoods and citywide coalitions will be elaborated on later in the discussion on the strategy of the struggle.

This vertical network is further extended outside gardens and endows the institution of community gardens with a more complex morphology. There are government agencies that oversee the gardens or own the lots on which the gardens reside. Most notably, New York City's Parks and Recreation Department (PRD) with about 400 gardens under its jurisdiction; the Department of Housing Preservation and Development (HPD) still holds about 40 gardens that are subject to development; and the Board of Education lists about 50 school gardens. Green Thumb, a federally funded program under the Parks and Recreation Department, oversees, licenses, and provides training to all gardens citywide and provides capital investment to Parks Department gardens. In addition, the New York Restoration Project (NYRP) and the Trust for Public Land (TPL)—which were described at length in previous chapters—purchased and saved a combined total of 126 gardens that were slated for auction by the Giuliani Administration.

Then, there are nongovernmental organizations (NGOs) that support the gardens in various ways. These organizations differ from each other in their political perceptions of the gardens and their role in the city. Green Guerillas is a nonprofit organization that grew out of the first garden in the city—Liz Christy, founded in 1973—and assumed the mission of helping other local groups in the process of creating and maintaining a community garden. Green Guerillas perceives neighborhood coalitions as a viable mechanism for gardens' sustainability and helps facilitate such coalitions. Another supporting organization is Neighborhood Open Space Coalition (NOSC) that used to assist gardens particularly with legal issues such as insurance.[22] The Council on the Environment of New York City (CENYC) is a semi-private organization under the canopy of the City Administration that provides gardeners with training, tools, and capital for small projects of infrastructure improvement. New Yorkers for Parks strongly advocates for more parks in the city and for a bigger budget to be allocated for parks and community gardens. Any budget achievements will benefit community gardens as well. There are about a dozen additional organizations in the city that provide some sort of support to the gardens.[23]

Many of NYCCGC's meetings include a guest speaker from one of these organizations (such as New Yorkers for Parks, Green Thumb, and Green Guerillas) who is invited to answer gardeners' questions. NYCCGC as well as neighborhood coalitions maintain relationships with these organizations but also with external organizations which are not part of the network of support to community gardens. These include organizations such as Slow Food, an international movement that supports local, artisan food vis-à-vis the culture of fast food; Community Support

---

22   NOSC stopped its assistance to the gardens in 2006 when the Parks Department announced that the City will carry the insurance for the gardens. NOSC was the only organization that refused to be interview for this work (for lack of time and resources).

23   I should mention at least by name and give credit to: Citizens for NYC, Trees New York, Brooklyn Botanical Garden, Cornell Cooperative Extension, Wave Hill, Earth Celebration, the Horticultural Society of New York.

Agriculture (CSA), which supports local farmers who struggle to survive in the face of agri-business; Just Food, which works locally to develop a sustainable food system in the NYC region; and Time's Up!—a direct-action environmental organization. Such organizations are invited to attend meetings and facilitate cooperation. Just Food, for example, is increasingly involved in helping garden coalitions initiate and operate farmers' markets and Time's Up! helps garden coalitions with direct action in regard to endangered gardens. In collaboration with Harlem United Gardens and More Gardens! they recently organized several bicycle rallies to protest against the assault on gardens in Harlem. Maintaining close collaboration with all these organizations is also a strategy to induce their involvement with the gardens' cause. Figure 6.1 recaps the structure of the institution of community gardens that was described above.

The discussion on the strategy of the struggle reveals more bluntly that the institution of community gardens in New York City is conscious of its mission and struggle. This consciousness entailed not only the major threat to the gardens at one point in history but also the gardens' ongoing existence as an alternative

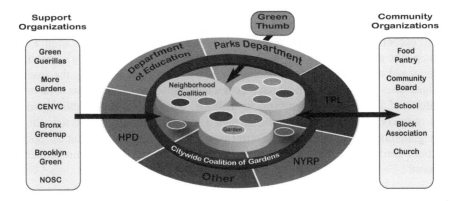

Figure 6.1   The structure of the institution of community gardens

space that is constantly marginalized by the dominate production of space. This consciousness is crucial for an alternative and contested organization of everyday life to become viable and long-lasting.[24] The strategy of the struggle of community gardens over the city—their resistance—stretches from grassroots praxis to institutionalization praxis. At the height of the struggle against the Giuliani Administration's plan to eradicate most of the gardens from the city, and up until 2002, a grassroots mode of operation dominated. This mode of operation included rallies, demonstrations, civil disobedience, the establishment of the bulldozer hotline, etc. that were initiated bottom-up. After 2002, with the gardens' settlement in place, an institutionalization mode became more dominant.

---

24   DeFilippis 2004.

Both modes of operation—grassroots and institutionalization—have been active at the same time, differing in dominance according to the historical context. I suggest three alternative categories of resistance praxis in order to overcome the above historical specificity of the gardeners' mode of operation over space. These categories encompass both grassroots practices and institutionalization practices, and are: "defining and demarcating space," "networking and communication," and "political power."

Defining and Demarcating Space

The most important strategy in the gardens' struggle is the appropriation and maintenance of space. The institution of community gardens strives to establish facts on the ground. The practices of taking over space and giving it life, character, and usage send out a strong message that the gardens are here to stay—they are needed and they are wanted. The successful protection of garden space requires the ongoing activation of several strategies.

The first strategy deals with the maintenance of a renewable membership in order to protect existing gardens. Outreaching new garden members is facilitated and encouraged in numerous ways. One example is proper communication with the public at large through signs on the gates that announce the garden's opening hours and extend the invitation to non-members to come in and enjoy the space. Signs also communicate necessary contact information for residents who are interested in becoming members. In addition, gardens organize public events to raise community interest in the garden. The events, usually in the form of a neighborhood party, are used to recruit new members and to raise funds. Sofia from a garden in the East Village provides an example of a membership outreach event:

> We do membership drives in June. We put [up] a big banner "membership drive—this weekend" and then we actually stand there with the membership papers and we try to lure people in, saying "You can become a member here and have unlimited access to it, whenever you want to come." We try to affect people from the immediate block, 6th and 7th Street but there are members who are from elsewhere also. And it is $20 for the year for a family. Not for a single person, $20 for the family. And you get put on an email list so you can be informed as to what happens in the garden and you get to hold parties in the gardens as a member and you get a combination for the lock.[25]

Recruiting new members becomes even more crucial in those gardens which become inactive. Such gardens are prone to lose their preservation status and get transferred to HPD. Neighborhood coalitions remain on the lookout for such

25   Interview, October 7, 2005, East Village, Manhattan.

gardens, report run-down gardens in their neighborhoods, and recruit, with the help of NYCCGC and supporting organizations, new membership. These actions cannot be done without a network of communication through which gardeners become aware of neighboring gardens and responsible for reporting problems to the coalitions.

Protecting endangered gardens is still part of the challenge for gardeners in the city. Old strategies such as rallies, camping in the garden, and demonstrations in City Hall (which now take place on a smaller scale) accompany newer strategies such as meeting with local politicians; developing alternative plans for areas that contain several endangered gardens, and presenting the plan to the community board and local officials; and activating the legislative process—making sure the ULURP is being followed by community boards and the city. The most recent battle over an endangered garden concerned Esperanza Nueva community garden in Harlem that was slated to be demolished to make way for a museum of African art and luxury apartment tower. The struggle involved rallies in Harlem and rallies and parties in and around the garden; numerous meetings with the local council member, Melissa Mark-Viverito; a presentation to the community board; several months of camping in the garden as a bulldozer-watch (during the winter as well), and more.

A press release from April 3, 2007 issued by a member of Time's Up! explains:

> All morning, as Nueva Esperanza Garden on East 110th Street at Fifth Avenue is being illegally destroyed by developers, NYPD helicopters are circling the area and the police are assembling off site, preparing to enter the garden to make arrests. Gardeners have chained themselves to the garden and climbed trees, stating their willingness to get arrested to save their garden. Environmental groups More Gardens! and Time's Up! have sent out calls to their members and volunteers to come to the garden to support, as press, gardeners, supporters and police continue to assemble. Gardeners who have been camping in woke this morning to the sound of the garden fences being destroyed by heavy earth equipment and trees being cut down by chainsaws. The encampment has been ongoing through the winter in order to watch the garden for just such an unannounced arrival. The city has essentially given away this land— probably one of the most valuable pieces of real estate in New York City—to the developers Sidney Fetner Associates and Brickman Associates for building luxury condominiums. This 80 million dollar project has been deceptively described as a "Museum of African Art" which has no collections and will only comprise a few small rooms for lectures. This land grab has happened outside of the competitive bidding process and without proper community, environmental or legal oversight.[26]

---

26   "Gardeners & Supporters Have Chained Themselves to Trees, Risking Arrest to Save Their Garden." Time's Up! press release, April 3, 2007.

The gardeners lost the battle and their garden but the group is still trying to secure an alternative space for gardening in East Harlem. They worked with many organizations during the struggle, most prominently the Harlem United Gardens Coalition, More Gardens!, and Time's Up! and also approached NYCCGC for help. An excerpt of the discussion on the garden in a NYCCGC meeting underscores the challenges that this garden group and the coalitions that support it face in the struggle over space:

> Benée Chisac of Nueva Esperanza Garden came to ask for the Coalition's help in resolving the issues with the loss of her community garden on 5th Avenue and 110th Street. Several relocation sites had been offered but unfortunately they have been now given to developers. Benée also pointed out that the current garden president, Anthony Bowman, is stepping down and that she will be acting as president. Ilya mentioned that all the previous sites that were offered were turned down by the gardeners. They wanted a site that was equally big to the site that they lost (future home of the Museum for African Art) and there were none around. He also said that it was hard to make a case for Esperanza as there was no paper trail and a year and a half had passed. Benée said she sent several letters to public officials and they have not been answered. She is now willing to accept the site on 119th Street and Madison. Billy mentioned the importance of showing a united front and speaking out as a group in terms of credibility. The Coalition will set up a meeting to mediate between the gardeners, but made it a point that it will be neutral in the process.[27]

Members of Nueva Esperanza are demoralized by the loss of their garden after an intensive year-long struggle; they have difficulty finding the collective energy to start anew. The institution of community gardens on its various components is also there to help this group strive and to protect it and other groups from falling apart.

The case of Nueva Esperanza reveals another strategy of the struggle which has been used by the institution of community gardens through the years—acts of civil disobedience. Acts of civil disobedience that have resulted in the arrest of many gardeners were used to raise public awareness and the reporting media for the cause of the gardens. The bulldozer hotline was established for that purpose, to call on gardeners to come and chain themselves to the gates and trees of a garden under threat from bulldozers. Ilya from More Gardens! talks about the use of civil disobedience in his coalition:

> The final thing is civil disobedience and direct action which is not always leading to lock-down in a community garden, but going out there and doing civil disobedience at HPD which we have done, we had lady-bugs locking down and climbing trees and so on at City Hall Park etc. etc.[28]

---

27   NYCCGC meeting, November 15, 2007, minutes.
28   Interview, May 31, 2005, South Bronx.

There are fewer available strategies for the creation of new gardens. In 2005 it was announced that all city-owned land was sold out or designated for development and henceforth there was seemingly no possibility of licensing spaces for new community gardens. However, NYCCGC and More Gardens! are advocating with HPD to release all the undevelopable sites for gardening. Ilya presents other ideas to recreate space for gardens:

> We are now coming towards [...] finding ways of creating more green space for schools and open spaces that are not necessarily buildable. There are still a lot of open spaces but they are grassy, parking lots, non-used roof tops, and we can move to the next level of opening community spaces that are in parks.[29]

On the same front, NYCCGC aspires to establish legislation that will include gardens in the zoning laws of new developments and—where possible—in existing built areas. The remarkably small ratio of open space per capita in New York City and the precedence of policy regarding community gardens in other US cities—such as Seattle's sustainable development plan that decrees the establishment of one community garden per 2,500 households[30]—help the legislative efforts of the coalition. A successful legislative outcome can guarantee the protection of all existing gardens and the inauguration of new ones.

Networking and Communication

Networking and communication represent other important strategies of the struggle. Since gardens are spread across the city and decisions regarding them are also made by municipal entities, keeping them connected and informed is a major challenge. The coalitions at the neighborhood and citywide levels constitute the network that enables the sharing of resources (funds, tools, knowledge, and human power), identifies needs and priorities, and responds to threats. The channels through which this network operates are mainly coalition meetings, annual gardener speak-out forums, newsletters, websites, and online working groups.

Steve Frillmann, the Executive Director of Green Guerillas, explains the arrangements of the network in the height of the struggle:

> In New York the importance of the coalitions is related to political changes that happened in the city in the late 1990s when gardeners were isolated from each other. It wasn't helping them to preserve their gardens for the future. By working in coalitions, they still fight for their garden with their peers but also helping each other out. It was the progression of time and New York City politics when

---

29   Ibid.

30   Lawson, L. 2005. *City Bountiful: A Century of Community Gardening in America.* Berkeley: University of California Press.

it become more of a necessity for gardens to interact with each other more, on some levels, not all. They are still independent, different from each other, run different programs. Somehow, the coalitions that we helped to start in the different neighborhoods were a political act to straighten the voice of the gardens in a time that they were assaulted by City Hall.[31]

But Sam from NYCCGC explains that the old arrangements no longer suffice, and ways of communicating ideas and orchestrating actions should go through the process of institutionalization:

> What happened in the past, there has been a number of coalitions like there is the Harlem United Gardens up here, and there is BANG in Brooklyn I don't know if they even still exist. But there have been different coalitions, each with its own particular agenda, and it is difficult sometimes consolidating all together and operating from one base so to speak or you know as uniformed basis that we can get.[32]

More creative approaches to communication are also part of a strategy to raise the awareness of the community at large of the issues of community gardens. Efforts to educate the public about the contributions and importance of community gardens are inspired by the idea of celebration (see Figure 6.2). Through acts of celebration people can connect with the beauty of the place and with its aesthetic, relaxing, and artistic elements. But through these presentations of attractiveness and playfulness participants are also exposed to new ideas and new sets of rules, and their curiosity about the complex social and spatial relations that are weaved through space is piqued. The "emphasis on elements of play, innovation, peer-based population education and respect for pleasure"[33] in efforts of social mobilization should be understood as "serious play."[34] It is a way to overcome the dominant tendency of individualizing social problems and cultivate "networks that allow actors to engage in new forms" of community organizing.[35]

Serious play is explicitly incorporated by gardeners as a means to advocate, publicize, and mobilize people around their vision. These activists specifically target the festive, carnival, and artistic arena of everyday life—the arena that has been mostly degraded in modern urban living[36]—in order to revive both the

---

31   Interview, September 29, 2004.

32   Interview, June 8, 2005, Harlem, Manhattan.

33   Shepard, B. 2005. "Play, Creativity, and the New Community Organizing." *Journal of Progressive Human Services*, 16(2): 47–69, p. 49.

34   Weissman (1990), ibid.

35   Ibid., p. 50.

36   On the disassociation of everyday life from its history and culture, and therefore its alienation, see Lefebvre, H. 2002. *Critique of Everyday Life: Foundations for a Sociology of the Everyday* (vol. 2). New York: Verso.

Figure 6.2   Educating the public through celebration

gardens and celebration in life. Ilya from More Gardens! describes this strategy of gaining public awareness and community involvement:

> We are going to do this in a very jubilant and colorful way, with puppets, with music, with singing. So we really show the joy and the celebration that happen naturally in most community gardens. There is always music and there is always somebody dancing, there are always kids playing, and there are always animals, maybe small, but there is definitely the party and singing and it all just happens naturally; nature sings with its roots.[37]

He continues to explain how this strategy works:

> Through the creativity and the visual aspect of it, people who would never come into the garden, who would never know about a demonstration, will come up and kids would come out and ask, "What is going on?" "What is this beautiful thing?" "Why is there such a parade?" You know, just naturally will come to that. [...] Just make that more wild and fun thing that people say "ho my god look at that butterfly, that huge butterfly" and there is a big sign on it "save the garden" or "make more gardens" or "less asthma more gardens" and all these things that people were signing, music old folk songs like "Gardens make us united" and "From east coast to west coast." So many wonderful creative people came together alongside gardeners as well as activists to make this united front. And it has been, I believe, very effective in getting a lot of media and at the same time not forgetting to educate people through puppetry, through everyday classrooms and bringing people together.[38]

The well-known annual Earth Day event (as celebrated in the East Village) is an example of this spirit of celebration integrated with the task of raising awareness of the issue of the gardens while referencing the city's political economy in general (see Figure 6.3). The Earth Day celebration is a colorful parade where gardeners march between the neighborhood's community gardens dressed as earth creatures, playing music, and blessing each of the gardens on their way. They also make stops to commemorate gardens that were bulldozed for redevelopment.

One of the highlights of the event happens when participants gather in one of the gardens for a joyful performance that narrates the story of the struggle against developers seeking to appropriate garden land. In this story the gardeners are portrayed as the devoted sons and daughters of Mother Nature and the earth goddess Gaia; the developers are represented as monstrous machines running after the gardeners and the audience and frightening them. And so the story goes on to tell how the earth creatures were able to oust the developers with their powerful magic powder extracted from their love of nature. This simplified myth-like

---

37   Interview, May 31, 2005, South Bronx.
38   Ibid.

Figure 6.3   Earth Day celebration

presentation of good and evil is very engaging and invigorating and the audience is informed of the plight of the gardens and their struggle. It demarcates the inner group (gardeners, city dwellers) and the outer group (the greedy and dangerous developers) and it is also a demonstration of power, representing the gardeners as the victorious ones. The event is fun and celebratory, yet delivers a strong message about the importance of the mission ahead (see Figure 6.4).

Another important strategy for networking and communicating with the wider public is the provision of various services to the community. By using the space of the gardens for food distribution to the poor, school nature classes, children's birthday parties, exhibitions of local art, and so forth, gardeners establish connections between local residents, neighborhood organizations, and the gardens. These services not only make the gardens known but also appreciated, valued, and cared for. Through this networking with local organizations and residents of the community, the gardens become part of the fabric of the community and mobilize more support and protection from the community.

Another component of the strategy of network and communication may be defined as the ultimate spatial communication mechanism—the map. Although most of the gardens are now under the jurisdiction of the Parks and Recreation Department, they are not included in the city map, and NYCCGC constantly approaches the city and the New York State Attorney General's office with appeals to map the gardens.

Figure 6.4   Performing the struggle during the Earth Day celebration

CENYC sponsors an online garden mapping system[39] which, after several years of lagging behind the changing geographies of community gardens in the city, was updated in 2008. These updated maps of community gardens include some new descriptive information that provides a clearer picture of the history and activities of each of the gardens. Nevertheless, official mapping of gardens on the city's map will grant gardens public status. Formally mapped park land cannot be removed

---

39   http://www.oasisnyc.net.

without state-level permission. Informally, but not less significant, objects on the map are documented in the cognitive perception of the space; they achieve a visibility which cannot be easily taken away. For the gardeners, "engraving" the gardens as green, named spots on the map signifies their ultimate victory after being dismissed as "vacant lot" for so many years by the city planners. More generally, locating the gardens on the map provides a visual representation of places of urban uneven development "so that they can be used by oppositional cultures and new social movements against the interests of capital as sites of resistance."[40] Mapping these sites of resistance and victory on the map is a tool of transforming the dominant representations of space.

Political Power

A major part of the strategy of the struggle of gardeners is to enhance their power within local urban politics in order to become a recognized and strong voice among official decision makers. Gardeners undertake three main strategies to enhance their political clout. They establish strong relations with political bodies within the locale; they develop relations with local officials and mobilize them to the causes of the gardens; and they construct the collective of gardeners as a valuable constituency for elected officials. Ilya from More Gardens! Coalition presents his political vision for the institution:

> Another [strategy] is working with legislative bodies governmental, community boards, etc. and one of our wishes is that as we build more and more of the coalition we start putting people into political power so that they can also really put that point out and say this is what we need these are the things. And not just be always asking but also say "Hey now we have the president of United States is from More Gardens! Coalition and we'll make lots of green space."[41]

The political body with a direct effect on gardens is the community board and its land use committee. Gardeners participate in the board and its committee's monthly meetings as a strategy to protect the gardens. A self-reporting survey of community gardens indicates that 50 percent (in 2006) and 40 percent (in 2007) of the gardens have a representative from the garden in their community board meetings. Participating in the community board is crucial for staying informed on the issues pertaining to the neighborhood and the gardens and being able to react. Ilya continues: "I myself am on community board 1, land use committee, so I'm always there and part of it is making sure that whatever comes on the land use we will know what it is"—and can act accordingly.

---

40   Pile, S. 1996. *The Body and the City: Psychoanalysis, Space and Subjectivity*. New York: Routledge, p. 3.

41   Interview, May 31, 2005, South Bronx.

Beyond monthly participation in community boards, gardeners attend Parks Advocacy days and other related events at City Hall, showing off their numbers and presenting their demands and concerns about community gardens and open space in general.

Gardeners establish relationships with local politicians such as district managers and council members to ensure their demands are heard. Neighborhood coalitions are particularly important for this purpose since they serve as a representative body

Figure 6.5   Panel discussions at the annual gardens' forum

on behalf of several gardens. Leaders of the coalitions initiate meetings with city officials, invite them to attend and give speeches at garden events, and politicians are invited and attend the annual gardens' forums (see Figure 6.5). Speaking to the participants of the NYCCGC meeting, Marvin from the Harlem United Gardens Coalition summarizes the achievement of gardeners on this front: "Eight years ago everyone thought that community gardeners are just gardeners, now we can see so many of the people here that their council member knows by name."[42]

Local politics has proved to be very important in saving many gardens through intervening in favor of the gardens. Being known by the community board and local officials created in many cases an open forum in which garden groups or coalitions suggested and negotiated plans for the neighborhood that fostered development while saving gardens, finding relocation sites, or proposing new open spaces for the area. For instance, More Gardens! Coalition, with several community gardens in Melrose neighborhood, Bronx, created an alternative development plan for the area, focusing on open space and gardens (see Figure 6.6). Ilya from More Gardens! describes the manner in which the dialogue on the plans for development was created:

> We went to Maria del Carmen Arroyo [Councilmember District 17th] when she got elected for the position and she was there listening to us and we talked to every other politician running for the same position and they said yes we are

---

42   NYCGC meeting minutes, May 18, 2005.

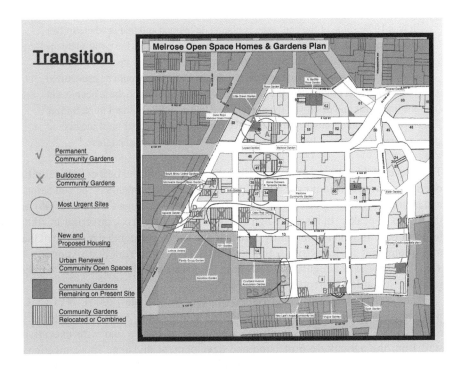

Figure 6.6   South Melrose, Bronx development plan

going to support the gardens. And she said yes, it sounds like a good plan, I will support it.[43]

In addition to fostering activities in which gardeners or gardens groups meet with influential urban actors, activists also make an effort to present the gardeners as a unified constituency that will support the candidate most responsive and favorable to their needs and demands. One way in which gardeners instill these messages is by showing up to council meetings and Parks Advocacy days with pins and banners saying "Gardeners vote!" The annual gardeners' forums to which politicians are invited as panelists and guests are part of the mechanism to show a united front of gardeners as a constituency.

However, local politics is obviously limited, being profoundly influenced by instrumental market interests. Gardeners, aware of their imbalanced power, direct their efforts also to bypass the local municipality and gain approval and support at state level. The intervention of State Attorney General Eliot Spitzer in the cause of the gardeners in 2000 stopped, through a court order, the destruction of gardens and helped in negotiations for a preservation settlement. Mike from a garden in the East Village describes:

---

43   Interview, May 31, 2005, South Bronx.

It was 1998, some of the gardens were being developed and that is when Spitzer got involved. I think it is when the Esperenza, where the Eastville Gardens building [now stands], built this huge fiberglass frog it was up on a platform, people camped out on it, so that if the bulldozers came they would be handcuffed to this frog as a protest. But the bulldozers came, they arrested these people, cut the handcuffs and destroyed the garden. And in the middle of that I think Spitzer filed a restraining order against all developments [of gardens] until they could determine if a proper land review process had been accomplished there.[44]

This successful intervention suggests that the state may be the best collaborator to overcome uneven local power. Eliot Spitzer became the gardeners' "patron saint" and honored them by appearing at the Second Annual Gardeners' Forum (2006) as the keynote speaker. In his campaign ads running for New York governor in 2006 he appeared sitting in a community garden in the Bronx surrounded by some of the most prominent garden activists. Spitzer's assistant as New York State Attorney— Tom Condon—was a regular participant in the gardeners' forums and was highly receptive to gardeners' requests and questions.

Once Spitzer was elected governor of New York State (2006), NYCCGC initiated a series of meetings at the governor's office. On April 2007 delegates of NYCCGC met in Albany with the governor's assistants and with executive staff from the agriculture protection and development division. These meetings were described by one participant as signaling another watershed in the history of community gardens in New York: a turning point of power structure in the city in favor of the gardens. Marvin from Harlem expresses his excitement:

As one of the six who made the trip to Albany, I want to congratulate everyone on their spirits, their presentations, their responsiveness, and our "togetherness/ unity" that was expressed in both of our meetings—with Governor Spitzer's special assistants, Tom and Paul and with the Dept of Agriculture's Director of Protection and Development and his assistant. History was made yesterday, and I feel very proud to say that I may have been a part of that moment. James, Ilya, Molly, Sam, and Rene—it was a pleasure being a part of the team yesterday. [...] History is about to be made in the world of community gardens in New York State. Go NYCCGC (HOLLA).[45]

A couple of months later the NYCCGC also met with Christine Grace, the manager of the Urban Food Systems Program from the New York State Department of Agriculture and Markets. The meetings yielded an agreement to collaborate on several issues. Among them is the production of a statewide forum for community gardeners; establishment of organizational support from the state; establishment

---

44 Interview. October, 5, 2006, East Village, Manhattan.
45 Reflection on the meetings at the governor's office in Albany, April 5, 2007 (internal correspondence).

of a process for applying to use state-owned land for community gardens; and the production of an in-depth brochure on the many assets of community gardens.[46] When Spitzer left his office in 2008, the NYCCGC board of directors and several coalition members also met with the new State Attorney, andrew M. Cuomo, in order to make their existence, needs, and problems visible.

Furthermore, NYCCGC recently discovered that in the past there had been a State Office of Community Gardens with an independent budget (not connected to Green Thumb and the city's Parks and Recreation Department). And so, since 2006 it has been working with state-level officials to reopen the office. It would like to make sure that gardeners will be directly involved in formulating the mission statement of this office and its operation.

All these efforts at the state and local scale towards strengthening the power of community gardens brought about several significant achievements that might represent, as Marvin suggested earlier, new prospects and a new era for community gardens in New York City and State. In May 2007 the Department of Agriculture and Markets opened a call for candidates to occupy the renewed New York State Office of Community Gardens. Representatives from the department participated in the Third Annual Gardeners' Forum, held in June 2007 at the Brooklyn Botanical Garden, and reported to the gardeners on this effort. Furthermore, in September 2007 the New York City Committee on Parks and Recreation (chaired by council member Helen Diane Foster) generated a resolution on "community gardens and operation Green Thumb" (No. 1033*). It was resolved that "The Council of the City of New York calls upon the Mayor of the City of New York to preserve the existing Green Thumb Community Gardens and to set aside more parkland, open space and vacant lots to be developed as permanent Green Thumb Community Gardens." The resolution was based on the assumption that community gardens are required in neighborhoods with less than 2.5 acres of open space per 1,000 persons. And in order to achieve this ratio, there is a "need to purchase lots through City, State, Federal and private resources to develop new Green Thumb Community Gardens, especially in neighborhoods with less than 2.5 acres of open space per 1,000 persons." The resolution included two other components that the coalition and activists had worked on for a long while. The first is the incorporation of community gardens as part of all new housing developments. In the second component Green Thumb is called on to "include provisions to protect the continuing existence of the garden, including provisions to map the garden as parkland or designate it as open space."[47] The legislative committee of NYCCGC "had worked hand in hand with Councilmember Helen Diane Foster's chief of staff Jim Fairbanks to bring about this Resolution."[48]

46   Meeting notes with Christine Grace, November 21, 2007 (internal correspondence).

47   Resolution no. 1033* New York City Council, September 24, 2007.

48   NYCCGC board of directors, November 21, 2007 (internal correspondence).

Grassroots vs. Institutionalization: The Tension

As the changes in structure and strategy of recent years indicate, the collective of community gardens in New York City is going through a process of institutionalization. The collective of gardeners consolidates the structure, missions, mode of operation, and spirit of its institution, and it has already made considerable progress towards establishing its position within local and state politics. There is, however, an inherent tension between the *grassroots* character of the garden coalitions and the various efforts to *institutionalize* community gardens within the space and politics of the city. The crux of the tension is the different and even contrasting rationales that underlie each of these modes of political operation. Simply put, grassroots action allows for innovative counter-hegemonic thinking and actions, unconventional and transformative demands, and a spontaneous, decentralized mode of operation. Institutionalization, in contrast, generates more conventional power, security, and stability but is required to adjust and to harness thinking and action to dominant ways of thinking and acting.

There are some obvious benefits to be gained from institutionalizing the collective of community gardens in the city. Institutionalization provides the *formal* tools that endow community gardens in New York City with power that exists beyond and regardless of the actions of individuals and groups within it. Institutionalizing the gardens has already led to the preservation of about 400 community gardens under the Parks Department gardens and established ULURP as mandatory for gardens designated for development. If the new legislation comes to pass, it will decree the required ratio of community gardens per square foot of a built environment, or per density of population. The power generated from institutionalization has already resulted in increased visibility for gardens, their struggle, and their value to the public among the general public as well as among political actors at the municipal and state levels.

However, there are limits and disadvantages to the power of institutionalization. The limits stem mainly from the social status of the project and mission of such institution. The manifesto of the institution includes issues such as sustainable planning, residents' participation, and civic agriculture. Though these may no longer be considered avant-garde, neither are they part of the mainstream: they do not compose the rationale that makes the city tick. The level of resources, funding, and political support that institutionalization can garner is therefore limited. The situation of the American Community Gardens Association (ACGA) is somewhat similar and improves our understanding of the institution of community gardens in New York City. A representative from CENYC explains:

> For so many years ACGA were way under the radar, because nobody knew what a community garden was unless you lived on a block with one. But now there is much more publicity about it, and not just the Giuliani publicity, but *Organic Gardening* every month has a whole page highlighting one garden around the country. So people are reading about this. A lot of other magazines have done,

there have been videos made, more and more research is being done so it has gotten somehow into the academia as well where it is a valid thing for scientists, mostly social scientists I guess, to look at and research. So now being on a higher visibility, do you make this leap where you become a more professional organization or do you try to keep that grassroots-ness about the organization? And I think there is a little bit of a tension in doing those two things. In some ways I don't think it will ever become a real professional organization like the American Medical Association [AMA] because even though I may be a professional greening person, the society doesn't really value that in the same way they value a doctor or a lawyer or psychologist, and that is fine, you know I'm not promoting that. But that also means that you don't get the same level of person in terms of wealth and influence that might be part of the organization that you would if you were part of the AMA. That same doctor is not going to join the ACGA or some lawyer that is in some legal association is not going to join the ACGA and bring with him or her the power and money they might have. So there is sort of a limit to growth in terms of the ACGA that way.[49]

However, it is important to note that the institution of community gardens in New York City, as probably is the ACGA, is a very different institution than these "high-end" institutions and associations. The latter organizations aim, among other things, to strengthen their position in the marketplace: they attempt to establish power which can be translated into exchange-value. The institution of community gardens is also acting to establish its power, but with an opposite aim: to de-commodify the space of the gardens and the labor that gets into it, as well as secure it out of market reach.

The disadvantages of becoming institutionalized rest in the loss of what might be called "grassroots-ness." In the case of community gardens in New York City this grassroots-ness is not only a mode of operation but also a part of their essence. The grassroots-ness and informality of the garden movement afford the spontaneous engagement with space and people, activate resourcefulness, and create an open forum for discussions and ideas and a cost-free space for arts, events, recreation, and celebration. They allow residents to participate in the spaces of their living environment in meaningful and transformative ways. Along with the power that comes from institutionalization, community gardens are also increasingly subject to discipline and oversight from external agents. These agents operated for several decades outside the realm of gardens and were detached from their causes. Institutionalization also means following guidelines and standardization which put limits on participation and artistic expressions of the residents and affects how particular cultures use the gardens. Ultimately, guidelines and standardization unify spaces and absorb their differences into the hegemonic matrix of urban spaces. The most salient evidence for this tendency is the new structures standardization guideline produce by Green Thumb for

---

49   Interview, December 5, 2005.

community gardens. After several decades of not enforcing any building standards whatsoever within the gardens—a non-policy that brought to existence the various unique structures that came to identify the gardens (such as the structure of the casita as well as many others)—the city is now going to dictate (and mostly forbid) the construction of structures in community gardens.

While the process of institutionalization is in motion, the spontaneous and decentralized structure of the collective (whose various components were described in previous parts of this book) is still being cultivated. TPL fiercely pushes for a full participatory model for the gardens by transferring legal ownership to the gardeners. Green Guerillas, the most prominent green organization supporting community gardens in the city,[50] is also guided by ideas of participatory democracy and established itself as a helper for the gardens according to their needs: an advocate rather than an organizer. As Claudia, a Green Guerillas field staff member, suggests: "I'm really going mostly to listen and to try to understand what the concerns are of the group and try to help the leaders of the group address the concerns of its members."[51] Her director at Green Guerillas recaps the organization's attitude:

> We perceive the gardeners we work with as neighborhood leaders. People who come out, walk out of their couch, went out to the street and actually did something, went out to the street and yell, wrote to the council member you got to fix that. In a lot of the gardens that you see it took so much to get to the point that it actually look like something. It is not an easy thing to do. So once you establish that you are kind of a leader that took a problem and solve it.[52]

In the same vein, Ilya from More Gardens! Coalition describes its mission as "only work[ing] together with the communities that *ask for us* or need our help to work together and build a community. *They already built a community*, we just accept that and work together with *their vision* of what they need and what needs to happen and then put up a resistance."[53]

At the same time, the participatory structure of the garden movement is challenged from outside as Green Thumb develops into a more imposing bureaucratic apparatus and NYRP functions as a private owner of the property it saved from the city. The participatory structure also faces challenges from within, manifested in the efforts of the coalitions to become more representative bodies and to unify (i.e. centralize) both agendas and actions.

---

50   Based on a self-reported survey of gardeners in 2007, 41 percent of community gardens indicated working with Green Guerillas, the highest percentage among support organizations.

51   Interview, November 8, 2005.

52   Interview, September 29, 2004.

53   Interview, May 31, 2005, South Bronx.

On this tension between grassroots and institutionalization, decentralization and centralization, it is suggested that "anarchist organizational form—the federation of independent groups which retained their autonomy—was most appropriate to a political formation that respected the tenets of participatory democracy."[54]

Activists are aware of the danger that institutionalization cast on their participatory model of operation. They realized the limitations of the agency that oversaw them—Operation Green Thumb—in the height of the struggle against the City Administration and the private market. With this realization they decided to break free and detach themselves from Green Thumb and establish a representative body of gardeners that can present a new, alternative voice unbounded by mainstream laws. These were the days when the movement of community gardens in the city was in formation. The first president of the NYCCGC, Sam, describes the formation of the coalition in 1997: "Green Thumb is a city organization, they could not really advocate for us. We had to find something that would speak to the issue so we began to have meetings."[55] Gardeners, then, could not depend upon Green Thumb if they strove to preserve their relative independence and their leverage of contestation, and put forth an oppositional agenda to that of the municipality.

Likewise, other supporting organizations are constrained by commitment to their funding resources and cannot support the more radical political agenda and actions that the gardeners like to pursue. Some organizations are constrained by their affiliation with the municipality and cannot assist gardeners with advocacy. Such is CENYC, which as a semi-public organization must adjust its agenda and messages:

> We can't, we pretty much don't advocate. *We can't really say something that opposes to what the mayor is saying.* So if the mayor has an environment policy that we do not agree with we really can't come out against it 'cause we are part of the mayor's office. It is also good that we are nonprofit organization because we have a little bit more leeway in how we can operate; city agencies are really very structured in terms of contracts and bidding and all kind of stuff like that. [...] We tend to be self-restraining. If somebody comes up with something that we think will not fly with City Hall we are not going to try to do it. And you know that has been a good policy because we have existed for 35 years and we [have] done good work, and we get foundations and individuals to contribute to our work, so I guess it is a good policy to have.[56]

The ambition of activists to remain independent is expressed in other ways as well. Gardens need capital in order to survive as affordable open spaces, and therefore

---

54 Aronowitz, S. 2006. "Is It Time for a New Radical Party? A Meditation on Left Political Organization." *Situations* 1(2): 117–58, p. 123.

55 Interview, June 8, 2005, Harlem, Manhattan.

56 Interview, December 5, 2005.

are always seeking funding. Gardens usually throw an annual fundraising party where the community comes and contributes. But gardens also seek grants from organizations that could potentially limit their action by enforcing their stance in decision making. Therefore, some activists are careful and consciously scrutinize potential donors to preserve their autonomy. Ilya from More Gardens! describes their selectivity in fundraising:

> We don't get a lot of big grants partially because the people who give big grants they have big hooks and claws and if we do anything that they don't like [...] we don't say, "Ho just cause you gave us $10,000 or $5,000 or whatever, we are not going to do it." We will be out there at your doorstep.[57]

Even in their relations with the dozen nonprofit supporting organizations gardeners strive to maintain their independency. They are concerned about losing their purpose and character if they become comfortable and too reliant upon outside help. At the 2005 Green Guerillas' self-evaluation session—responding to the question "What can Green Guerillas do for the gardens?"—Rene, a gardener from the Bronx and a long-time activist at NYCCGC, pleaded: "Help us become self-sufficient, give us the tools to help ourselves. We should not become depended on greening groups, it makes us lazy. We should be able to get along without you [...]."[58]

Gardeners, then, seek a constellation in which their autonomy could be preserved and their agenda could be voiced unconstrained. Critical theorists suggest that the place of real autonomy, and therefore the only place one can find the "No!" is in the margins. For Adorno and Horkheimer of the Frankfurt School, for instance, the margins are art and myth. For Lefebvre and Harvey the margins— the places of potential contestation—are on the capitalist map, where capitalism chooses not to invest and not to develop. Community gardens grew out of uneven urban development in the forgotten places of capitalism; outside the playground of the hegemonic forces and therefore not implicated by their instrumental rationality. Gardeners realize their marginality as well as its potency. It is there, in the margins, that gardeners found the space to say "No! We want something else for ourselves and for our city and we can do it differently." However, with the acceptance and power that institutionalization entails, gardeners move from the margins where a real critical positions can sprout, closer to the center where such positions are harder to espouse.

Claudia, a gardener from Harlem and a staff member of Green Guerillas, can already observe the purging of some of the unique capacities and character of community gardeners as a result of their increased institutionalization. She observes a loss of the spirit that makes things differently, in addition to increased dependency and compliance. She asks:

---

57   Interview, May 31, 2005.
58   Activist, Green Guerillas self-evaluation process, December 8, 2005.

What are those gardeners entitled to from either Parks or the Land Trusts? It seems that when community gardens started in NY there wasn't that much sense of entitlement, in the early 1970s. People just took over land and used all the resources that they could. And now I think there is sort of a sense of many gardeners that they are entitled to services from the city or from the land trust. And I think it is kind of tricky. Yes I think the city should support the gardens when they need it in terms of basic materials. And yes, I think the land trust should provide material to their gardens. But sometimes when I'm talking to gardeners I feel like I want to remind them of the first period of community gardening and say "Be resourceful. You don't have to wait for years for Green Thumb to bring you these bricks or whatever. Go around the block to the construction site and see if they have materials they are throwing away. Do something besides waiting for the city to bring services, that is not really getting you anywhere." I don't know how it happened that that sort of mentality got cultivated among gardeners that they are entitled to city services. I think that what was exciting about community gardens and even for me now personally is being resourceful and make do with what you have around you and make something really beautiful out an abandoned lot in a way that is simple, that it doesn't take a hundred of thousands of dollars.[59]

This "mentality" of gardeners that Claudia bemoans developed as a result of the change in the position of power of gardens in the city and the gradual mainstreaming of the gardens. The institution of community gardens emerged out of the desire to release gardeners from the mainstream political apparatus and establish them as a counter force fighting for an alternative kind of space. The success of the institution also doomed it to walk the mainstream path.

Does this tension between grassroots and institutionalization exist in all counter-forces? Does institutionalization entail the loss of capacity to say "No!" and the coming to terms with the unavoidable mainstream path? Can a counter-force, such as a collective of community gardens, induce a real change in spatial and social relations without institutionalizing? The answer to of all these important questions might be beyond the scope of this book. However, the ways that the institution of community gardens in New York City is dealing with this tension can at least shed some light.

The core components of the institution—NYCCGC and neighborhood coalitions—are pushing for further institutionalization of community gardens in the city. As already mentioned, they work towards a more comprehensive legislation for community gardens, on a state-level system of control (the State Office for Community Gardens); greater access to city-level decisions in regard to gardens (the Green Thumb advisory committee); higher visibility through mapping gardens as park land; and in general towards increasing their representability among gardeners and outsiders. At the same time, they act to invigorate gardeners

---

59   Interview, June 8, 2005, Harlem, Manhattan.

and reunite them under the causes of the gardens rather than being comfortable and atomized (resting on their laurels). They use various means to achieve this, such as calling for vigilance from new threats and reminding gardeners that the battle for the future of the gardens is unknown and demands preparation and action (most notably when the expiration date of the settlement in 2010 was approaching). Sam from NYCCGC explains:

> I think the main thing is trying to *keep the energy*, trying to *keep the momentum going* in terms of the existing gardens. There are so many people of the gardens that are under TPL, NYRP or Parks and *people are not coming out like they did when the gardens were under immediate threat*. People feel like "my garden is saved." But [...] *we have to be vigilant* and pay attention to what is going on from now on. A certain administration can come in and they can decide that gardens are just worthless they might try again to take the gardens. As it is now we are in a pretty good shape as it stands now, but who knows what can happen in the years to come. So we need something for the long run. *We need an organization like the coalition to pay attention* to that and to educate the public on the importance of that. And not just being gardeners, *we are also guardians* of the land so it is really important to pay attention and stay in focus. Because you know how it is in the city [...] things can change legally and laws change all the time. We need to study and keep ourselves at breath with the issue and watch the land use and what happen to the land and really need to maintain our gardens.[60]

The coalitions use invigorating slogans that worked in the past to entice people to come to the meetings, the annual forums, and to stay focused on the main issues. In their efforts to keep the discussion alive they ask gardeners "Is your garden really safe?," "What are the main challenges that your garden faces?" They also constantly approach the administrative branch with inquiries about the end of the settlement, the future of gardens in the city, and inquire about scenarios in which a different city administration foresees different futures for the gardens.

The City Administration, being dismissive of this tone and wishing that gardeners would go back to being just gardeners, tries to communicate that the struggle is over. But at the same time they cannot ignore the actions of the coalitions. Administration representatives show up at the annual forums and reply to the questions; they are aware of the progress gardeners made at the state level; and recently they allowed the formation of an advisory committee (of gardeners) to Green Thumb. This committee will have more access to issues and decisions at administrative level than gardeners have had before, but it might also be understood as a means to eliminate some of the critical edge of the coalition. A Green Thumb representative criticizes the coalitions' tone of continuing the struggle:

60   Ibid.

The gardeners are very loath to give up the struggle, you know, they like the struggle, the struggle kind of defined them for a while, so there are certain people that were involved in the preservation fight that will never believe that it is really Parks land and that they are really safe. No matter how many times we said it has a Parks property number the same as any other park, "it is not mapped though." OK, it is not mapped but neither is a lot of Park land that the city has. The city doesn't map park land that is given to them from another city agency; it is more paperwork than it's worth. "But then it can be taken away." Ho, now Spitzer is the governor, I guess that makes it slightly safer but in the past, all you have to do to de-map any park land was to get an act of a legislator which really wasn't very hard to do; it isn't really that much more protected. There are some certain groups that always want to consider themselves as the adversary. They want to be the counter-culture. I don't know I think that is part of it. It is not everybody, not all the gardens, some gardeners are just happy now that all they have to do is garden.[61]

While dismissing the spirit of the struggle as no longer relevant, this representative provides—perhaps unintentionally—the justification for its continuation. It was already mentioned that the city pledges the long-term preservation of Parks Department gardens based on the complicated process of "de-parking" park land. The assistant to the Parks Commissioner flagged, in two consecutive gardeners' forums, the difficult procedure of de-parking in order to rebut concerns regarding the future of the gardens. At the same time, the above quote from 2006 dismisses the strength of this procedure as a bulwark against the destruction of the gardens. In light of this, activists' conclusions that the municipality is unreliable and that the city is unpredictable seem valid and justified.

But preservation, as discussed at length in this chapter, takes its toll. This new status of gardens affects the autonomy of gardeners and of the gardens as grassroots independent units. The relations between the gardens and the organizations that oversee them (that used to be loose and temporal) have to be clearly determined now. These defined relations mean regulating the movement: rules of operation, buildings standards, etc. The Green Thumb representative continues:

Now we are sort of turning, we and TPL and the other groups that are organizing are making rules, now we are the bad guys. "Why do we have to do this now, why are you making us do this, why don't you just let us do what we want to do, it is our land?" Yes it is but at the same time it is a bureaucracy and you're gonna have to deal with that. And I feel that completely myself, I have been out as an activist, now I'm a bureaucrat. I get bored now easily. And pretty soon I'm going to leave, once all the stuff with HPD is over, running Green Thumb is going to be more and more about enforcing standards and that is not me, that is not my thing. Going around and saying, "Ho, these guys are having a roster,

---

61　Interview, November 15, 2006.

make sure they get rid of it." That is not my thing [...]. The building department did finally make regulations so that is another thing that will happen in the next year or so, that we actually do start enforcing building department regulation. Can you imagine how much the gardeners are going to hate that? They are going to be very angry.[62]

These examples uncover an interesting paradox that questions the overall ability of the local government to embrace such differential spaces as community gardens. In the past, the city endangered the existence of the gardens by favoring market interests and ignoring the meaning and value of the gardens to local residents. At present, although the city and the institution of gardens arrived at an agreement that favors collaboration over confrontation, the city threatens the autonomy and freedom of the gardeners—this time as a bureaucratic apparatus. That is, whether deploying the language of the market as its rationale, or the language of bureaucracy and administration, the underlying motivation of the city seems to be the same—to control differential space and to colonize it into the prevailing conception of space.

Community gardens, as a type of collective ownership and management of space, need to obtain local autonomy in order to facilitate social transformation. We have underscored the mechanisms with which the institution of community gardens appropriates and influences the local institutions of everyday life. We have also seen how the institution of community gardens develops and operates with reflexivity over its internal conflicts, central of which is between institutionalization and grassroots-ness. In efforts to protect itself from the mainstream effects of institutionalization, the institution of community gardens develops, preserves, and presents oppositional politics and reflexive consciousness of the struggle among gardeners. The institution of community gardens epitomizes transformative urban action that succeeded in translating everyday social practices on space into political action and discourse and allowing people with varying degrees of political awareness to find a more politically involved role in their community. If the importance of movements lies in helping people to see and direct their attention[63] then the institution of community gardens does exactly that. The kind of "direction-ness" that the institution of community gardens presents is unique because it is presented not only as an idea but it is also coupled with praxis that is engraved in space.

---

62   Ibid.
63   As suggested by Aronowitz (2006).

Chapter 7
Setting the Ground for "Organic Residents"

What is the broader impact of the production of such spaces, such associations, such actions and discourse on urban residents who participate in this socio-spatial assemblage? Part I suggested that the different opportunities encompassed in this assemblage afford residents means to become more resilient, capable of dealing with the vagaries of urban life. Part II discussed the modifications and adjustments of community life that grow out of this socio-spatial assemblage. Participants reworked their living conditions to help compensate for the uneven distribution of urban resources and in the process became more significant social actors. But this assemblage of spatial practices, social structures, and collective wisdom is responsible for an additional contribution. It gives rise to a dialectical force in which the evolving institution of gardens serves as a breeding ground for personal political development and vice versa. This chapter alludes to the broader, self-generated relations between the institution of community gardens and opportunities to engage in political practices (in the broad sense of the word) as well as opportunities to take part in the process of production of space. The self-generative power (autopoiesis) of this assemblage endows it with a life of its own; this power promises a generation of new participants and sustainability of the institution in the long run. The development of political consciousness remakes participants as a different kind of urban resident; to paraphrase Gramsci, it produces *organic residents*.

Before elaborating on this new concept let us draw and demarcate the process by which political consciousness is generated. It is most prominent in the experiences of residents who join a garden without having pretense to execute a political agenda but for personal reasons (those gardeners who were classified in Part I as presenting the "reactive narrative"). Once members, though, they are "picked up" by the energy of the collective—propelled by leading figures, by meaningful practices over the space that affect their life in important ways, and by the discourse that lends new understandings and new visions. They develop new practices which are coupled with new understandings; they learn things that prompt new practices; and this construction of practices and discourse becomes their new political consciousness.

Political consciousness, or the process of *concretization* (à la, Freire), percolates through the various channels of spatial practices and social interactions. Individual gardeners are exposed to discussions on socio-urban issues in membership and coalition meetings, in celebrations (recall the performance in the Earth Day celebration), and in public events such as Green Thumb's GrowTogether or the annual gardeners' forums. This discourse is also present in mundane and spontaneous interactions in the garden, which thereby becomes a perpetual arena for discussions that go beyond the particularities of growing plants.

The structure of the collective of gardeners—the relations between the different units and the way each unit functions—generates a momentum that pulls participants into action and constitutes them as gardener-activists. Joshua, who joined a garden in the East Village because he wanted to take care of something, to grow something (reactive narrative), tells how he later became a board member in charge of the landscape committee:

> They asked me to become one just because I seemed to have more interest in it than a lot of other people and the spot was open and I guess I proved that when I am set to do something I have try to see it to the end. So they appealed to me and it was also interesting to me because I came here for the garden but then it is just a very political environment [...]. So you just start getting involved in the politics here and the politics of the community and the board, the board is kind of monthly meetings when people talk about things that are going on in the neighborhood, maybe you can help, maybe you can't but it is good to know what is on people's minds that live here, and I have lived here for a long time.[1]

Joshua wryly presents how he became a board member: a combination of needs of the board and his own potent qualities. But the significant part of Joshua's experience came after he actually stepped into his board position: the possibility to experience the neighborhood in new ways, to become knowledgeable about the important forces that shape his living environment; to get involved in discussing them and to practice control through this involvement, whether it materializes in actual influence or not. By getting involved and participating Joshua not only finds the connection to the "pulse" of the place he lives in but he also becomes a producer of knowledge (local knowledge) and of space.

Emily, who grew up in rural Oregon, joined another garden in the East Village to satisfy her need for green space and a sense of belonging and a sense of home in the city. However, her involvement in the garden was quickly extended beyond experiencing and caring for nature:

> I'm the vice president of the garden, just recently elected, this is my third time since I've been a member on the executive committee and, you know, this garden is lovely, it is like my backyard, it is where I raised my child. Her and I we come in here year-around, summer, winter, fall, spring, we are here, regardless of the weather. It is very political, so there is a lot of stress involved, a lot of people who you have to just compromise with, there is a lot of compromise that goes on all the time. We had at one time a lot of outside politics. There was a time when the city wanted to build here; they wanted to take it away. We had a lot of battles over how to save the garden. And then there is a lot of inside politics. It is very

---

1   Interview, September 27, 2006, East village, Manhattan.

meaningful to me. I participate in a lot of the activities that go on [...] so I'm very involved in that way and that is really great for me [...].[2]

According to Emily, her involvement in the garden is not only about having fun but also requires a lot of negotiation and compromise. At the same time, this more complex and complicated character of experience makes it meaningful and worthy. Taking part, even in the arguments, makes participation meaningful. Emily has a say on matters; she can fight for or compromise on her ideas. Gardeners' participation, moreover, is not only about voicing ideas but also materializing them in actual results; a change in the space, the experience of it, and its operation.

Support organizations whose mission has to do with mobilizing gardeners' participation—for example, Green Guerillas and the New York City Community Gardens Coalition (NYCCGC)—utilize the enrapture power of political praxis. They attract gardeners' attention with things that are close to their heart: plants/ seeds giveaways, important information on the operation of gardens, and the like, and through these other issues lure them into action. Claudia from Green Guerillas demonstrates this process through an action initiated by her organization in Bedford-Stuyvesant in Brooklyn that was intended to foster a better organized garden coalition in the neighborhood: "We had mailed out about cover crop seeds and that we have cover crop seeds for free and someone [from the garden] called in about them and then we went and met with them and they were excited about collaborating on all kinds of other projects."[3]

Since many gardeners joined community gardens for reasons other than political, they did not typically expect to deal with the complexities of the world they entered. Billy, a savvy warrior against threats to his garden in downtown Brooklyn, took part in the struggle against Mayor Giuliani's intention to evacuate his garden and then again had to fight against plans to destroy his garden in favor of the Forest City Ratner Atlantic Yard development project:

> We were forced into [the struggle] when Giuliani tried to sell the gardens and suddenly we became much more versed in civics, in city government and in local government than we wanted to be. We did not want to have to learn to talk in front of city council, or to try to save our garden, or get arrested in Manhattan for being a gardener. You know what I mean? It is not something that we set out to do when we first joined community gardens. We just thought we were doing a really nice thing: a greening project in the city that the city desperately needed.[4]

There is discord between the original intentions of gardeners and the reality they have to face once the space they produced (and become highly attached to) contests, negates, or clashes with the dominant space.

---

2   Interview, April 24, 2002, East Village, Manhattan.
3   Interview, November 8, 2005.
4   Interview, July 29, 2006, Crown Heights, Brooklyn.

Mr Thomas from East New York, Brooklyn, who joined the garden for the green space and fresh produce, describes the experience of being picked up by a wave of participation and action:

> Slowly I began to learn a little more because I'm not a gardener at heart. I don't know anything about this, I'm concrete. I'm the city guy. I do love fresh fruits and vegetables and I love the earth. I love the greenery. I love the quiet like what we are doing now, I can do it forever. And I began to learn from other gardeners and you got the Green Thumb, the Green Guerillas and the Hannahs and Rebeccas [names of two prominent staff members of Green Guerillas]. And there is a whole world within the greening community which I had no idea about when I first met my Christian friend with his garden. I was thinking, "Oh we got all the space, I can get cucumber ..." No, it is much bigger than that. So slowly I got involved [...].[5]

The land of opportunity, literally and figuratively, was revealed to Mr Thomas and like-minded people upon joining the garden: land that provided a joyful urban experience but also comprised of groups' efforts to maintain it as an open environment for participation—in ideas as well as in action.

With an enriched awareness of the politics of community gardens, Mr Thomas became an active member of the East New York Gardens Association. He is a self-invented resources-coordinator for several gardens in the area and takes part in citywide gardens' events and coalition meetings. Billy, as another example, played a major part in the Brooklyn coalition's struggle against Mayor Giuliani's auction plans and then in the successful fight to save his garden from Ratner's Atlantic Yard plan and secure it as a Land Trust garden. He took part in Brooklyn-Queens Land Trust organization efforts, was a member of the group that initiated the First Annual Gardeners' Forum, and participated in the planning of subsequent forums. He was also elected to the NYCCGC board of directors in December of 2007. Billy is only one example of gardeners who experienced and accumulated an impressive record of action and accomplishments. The institution of community gardens, with its different units and branches, encourages gardeners to develop as participants and activists.

Sofia from the East Village represents another example of how spatial practices can evolve into political practices with well-developed political consciousness. Sofia became a gardener because of an inspiring leading figure—Françoise Cachelin. And she was the one who pulled her into action:

> [...] that is interesting because I was never involved in that so much. When Françoise was there I was helping her; I didn't have a title. I would just go to City Hall when they had hearings. I would go when we had a bulldozer [alert], things like that. I would help stuff envelopes. I would make flyers for her if

---

5   Interview, May 6, 2006, East New York, Brooklyn.

we had rallies. But I was never involved in the organization. I went to some meetings when we plan Standing Our Ground. That is how I met Sam and Rene.[6]

Thus Sofia has done a lot to help the movement but felt that she was "on the sidelines," helping Françoise with whatever she had asked of her. But Sofia did not stay on the sidelines for long; her practices had their own momentum and the institution of community gardens has various mechanisms to pull participants into action:

[…] just one day out of the blue Sam called me and he said, "I know you were very close to Françoise and I know when we had the memorial we all pledged to continue what she had started, to not let it die as sort of the best gift we can give to her. And I don't want the coalition to die. It sort of went dormant after she left and she was such a pillar of it. And I know you were very close to her and we would like to ask you to join the coalition." So we meet once a month. That is sort of how I started getting involved.[7]

Sofia started as a member of the coalition, not missing any of the meetings since May 2005, contributing to the revival of the institution that Françoise, her friend and mentor, had established, and learning about it from within. After several months Sofia joined the NYCCGC board of directors and since then she has been at the forefront as one of the young spirits that propels the organization.

At the forefront of this assemblage are inspiring people and visions. Some of these people may better be indentified with a proactive narrative of joining a community garden. Unlike Sofia, Billy, and Mr Thomas who presented a more reactive narrative, they joined in, as a representative from the Council on the Environment of New York City (CENYC) describes, because they saw the potentialities of space to generate a broader impact. The problem for this group, he suggests, is their limited number:

I think there are some gardeners who are very politically savvy, that look at the garden as a way to organize people in their community and in a larger scale as well. And I think a lot of those folks, the people that you might know from the garden coalition, they see that they can't just garden and be happy planting seeds and pulling weeds and that's it, but that there is more to it. So I think there is a number of people, and it is not a large number either, that also see the garden as kind of a stepping stone to greater community participation and greater political participation as well.[8]

---

6   Interview, October 7, 2005, East Village, Manhattan.
7   Ibid.
8   Interview, December 5, 2005.

Nevertheless, this observation on the limited number of politically savvy gardeners overlooks two major components. First, that the formation of this political consciousness is a *process* and second, that this process has very powerful *generators*.

The politicization of gardeners, the process of the formation and development of political consciousness, occurs on several scales. There is the scale of the garden unit that branches out to the neighborhood (as exemplified in the stories of Joshua and Emily); the scale of the neighborhood (coalition) that connects with the citywide level and with other neighborhood coalitions and organizations (as in the case of Mr Thomas). And there is the scale of the citywide coalition that branches out to other city-level organizations and agencies and to the national level, as well as feeding back in and mobilizing communities. The structure of the institution affords participants to move between the levels and reduce or intensify the extent of their political practices, but still maintain the political consciousness that has developed.

As with every process, the process of forming political consciousness also has a temporal context. Different historical periods influence the process, slow it down or invigorate it. The struggle against the Giuliani Administration, for example, was a crowning moment that fostered and fueled the process. The political practices yielded successful results that in turn slowed down the process of political development, but did not eliminate it. Yet again, since 2004, the process was picked up with renewed and slightly different energy. The metaphor that a representative from CENYC used to explain why the future of community gardens was not in danger can be assumed to characterize those changing moments of the process:

> I like to use this organic analogy of the soil, or as organic gardener, of a compost pile. It goes through several phases. In the case of a compost pile if you're trying to do a quick compost where it heats up high 'cause there is a lot of activity, it is like you are just creating something new and people are meeting and lots of things are going on. And then it is going to cool down for a while. It could be for a whole number of reasons. But then if you turn the pile, it will heat up again the second time and eventually that compost, as it breaks down, is going to become the seed for something else. It is something that is growing.[9]

This metaphor alludes to the powerful generators that help sustain the process of politicization, which are: 1) the vision, 2) the institution, and 3) the space. Each by itself, and more so together, fertilizes the development of political consciousness and protects the process from withering.

Activists, especially those at the forefront of the movement, present a clear *vision* of the city that they would like to live in, a vision that is coupled, in their perception, with a strong sense of responsibility and mission. They perceive

---

9   Interview, December 5, 2005.

themselves, in the words of Sam from NYCCGC, "not only [as] gardeners but [as] the guardians of the land."[10] These savvy gardeners promote a vision of a fully-fledged transformation of the city according to which its space becomes more green than gray; its circulation mechanisms of food, products, water, and garbage become localized and self-sufficient; and the users of the environment have a powerful role in producing it. Their urban vision may paint a picture much more grandiose than what can be achieved and materialized, but with it they draw the utopia and create a direction for action. Moreover, they identify community gardens as the primary bridge (an actual bridge) to progress towards this vision. The actual space of the gardens is their blueprint for a political strategy. Their vision is not day-dreaming but a conscious move based on the belief that if they fight for the "whole" they will eventually be able to get more of its "parts." The more concrete version, though, includes alleviation of the inequitable distribution of residents' access to green open space and recreational opportunity, allowing people a sense of ownership over a piece of land and an opportunity to be involved in its production and, more generally, an augmentation of people's involvement in the production of the urban environment.

The *institution* interlinks the vision with the space. The institution of community gardens aspires to grow in number and to broaden in scale. The actions of the institution—connecting with other organizations, establishing a recognized identity, and increasing levels of visibility and representability—are also means of articulating ideas and developing the consciousness that the group holds about itself and the world. The links that the institution of community gardens established with greening and environmental organizations—organizations that contest commercialized cultures of consumption such as Slow Food, and even local community organizations such as soup kitchens and schools—inform the vision of gardeners and their political consciousness. The institution of community gardens and its ideology are mutually and continuously forming. In that sense, this complex network might be thought of in the Gramscian sense, as the *institution* that in the process of its production both reveals and constitutes political consciousness.[11]

The transformative potential of the institution of community gardens in New York City can be attributed to five of its characteristics, five factors that are important in activating residents and developing their political engagement:

1. There are multiple roles that individuals can take up within the institution.
2. The different roles are not in conflict with each other.
3. Roles are not binding—participation is easily kicked off or withdrawn.
4. The process of institutionalization is dynamic, thereby creating new roles and the opportunity for people to reinvent themselves.

10   Interview, June 8, 2005, Harlem, Manhattan.
11   DeFilippis J. 2004. *Unmaking Goliath: Community Control in the Face of Global Capital*. New York: Routledge.

5. The material core of the institution—the gardens—underlies all roles and functions; therefore no matter how broad or abstract the practices of the institution may be, it is connected to the material space of the garden and the spatial practices that are part of individuals' everyday life.

Broadly stated, it is the conflation of everyday spatial practices and a broader political discourse that facilitates personal and collective political development. Gardeners might participate in meetings and events for very practical reasons (for example, having trouble with covering insurance premiums), but in the process they become familiar with a political vision and assume the role of activists. When doing so they are not simply assuming a political identity but becoming part of a collective, part of an institution that they can identify with and be represented by. The *community* of community gardens that both contributes to and criticizes the surrounding urban environment is the kind of community that can serve as a real critical buffer between the state and the market.[12] Gardeners, attached to their collectively produced environment and identifying with it, develop their identity as meaningful participants in society through their relations with their community.

Last but not least, *space* is a major generator of the political consciousness of gardeners. Space entails spatial practices, and spatial practices of the type that can be found in community gardens are destined to be transformed into political practices. Space has an electrifying capacity to pull people into action and to politicize them. This is because space is never neutral: it embodies social conflict and power relations, and these are revealed and challenged through the social production of space.[13] The collective action of people over space, in which people have the opportunity to use space according to their needs and to participate in its production, situates them as stakeholders. As stakeholders these residents obtain a new powerful status within the urban constellation of power; and while being active over space, this status further politicizes them. People who use space in a way that was given to them, for instance as spaces of consumption, do not reproduce it for themselves (according to their needs and visions) and do not intervene in the power relations that produced it. The common acceptance of space as a commodity, in accord with the hegemonic property regime,[14] is reifying space and further abstractifying and flattening it to this sole meaning: abstractifying and flattening because exchange-value of space is devoid of all the social, cultural, and spatial practices and meanings that are outside the practice of commerce. The multiple meanings of space may be accentuated through appropriating space for its use-value and reproducing new designations and meanings to it. Political consciousness springs out of new forms of using space and being in space and

---

12   Rose, N. 1999. *Powers of Freedom: Reframing Political Thought.* Cambridge: Cambridge University Press.

13   Lefebvre, H. 1991. *The Production of Space.* Oxford: Blackwell.

14   Blomley, N. 2004. *Unsettling the City: Urban Land and the Politics of Property.* New York: Routledge.

the new relations and values that are generated from it. Furthermore, the promise of more power, more control, greater equity, and better environments that is encapsulated in the alternative production of space is an important catalyst for politicization. Thus, following Lefebvre, social production of space is viewed as a means to create new forms of politics and to change social relations.

The institutionalization of community gardens in New York and the development of politically conscious participants amount to *resistance*. Resistance comes about when reworking practices are "armed with" oppositional consciousness. Oppositional consciousness entails the recognition of oppression and marginalization, coupled with "a vision of what else could be." Practices of resistance, viewed in this way, seek to redress the "imbalances of power and resources."[15] It is the development of reworking practices into a broader consciousness and a broader perception of the problem, beyond its local occurrence. For many years community gardeners benefited from their interactions in and with the gardens in multiple ways. They were able to de-alienate their living environment, to identify with it and to become more resilient in dealing with their urban condition. They were able to rework and improve their lives in the city through the garden. Gardeners had the experience (not only the vision) of what else could be done in order to make a different living in the city. When their gardens came under threat, the position of gardens within urban political economy and their own disadvantages within the urban power scheme became apparent to them. They therefore developed oppositional consciousness and oppositional practices to redress these constellations of power.

The binding of oppositional political consciousness with everyday garden activities and engagements forges what gardeners call "sparks of new energy." Therefore, the gardens should be understood not simply as the *end* of resistance but as the *means* to redress other injustices. For example, the gardens serve as a constructive arena to deal with issues of racial misrecognition coupled with spatial maldistribution[16] as these are epitomized in urban ghettos such as the Bedford-Stuyvesant neighborhood in Brooklyn. One of the interviewees tells of:

> a young, early 30s black guy from Bed-Sty who grew up in Bed-Sty who is really sort of thinking about what are the important issues in this neighborhood and what can you actually do to change those things. He talks a lot about positive-male-black role model and stuff like that. And that relates him to getting involved in the garden and getting youth from the neighborhood involved.[17]

---

15   Scott (1985) in Katz, C. 2004. *Growing Up Global: Economic Restructuring and Children's Everyday Lives.* Minneapolis: University of Minnesota Press, p. 253.

16   Fraser, N. and A. Honneth. 2003. *Redistribution or Recognition? A Political-Philosophical Exchange.* New York: Verso. "Misrecognition" and "maldistribution" are Fraser's definitions of status and class injustices respectively.

17   Interview, July 2, 2006.

Being uniquely positioned—presenting both a microcosm of social relations and a deviation from this cosmos (both a space of reproduction and a differential space)— the gardens provide a fertile ground to think about social realities of inequality reflexively.

Another example of the garden as a space of resistance, in Katz's term, is the construction of oppositional ideology and practices to neoliberal global culture. Here an alternative mode of living is not only envisioned but an effort is also made to realize it through the space and practices of the gardens: "The Time's Up! group, like the bicycling group, they really went instrumental in getting a couple of gardens going in Brooklyn. [...] They kind of have to do with the idea of being self-sufficient and not be plugged into this oppressive political consumer culture system."[18] Practices, like in these examples, that are afforded by the gardens are opportunities to redress many of the pivots of modern urban life. The production of differential space ignites sparks of oppositional political consciousness and the production of the institution keeps the fire going.

Conclusion: What Are Organic Residents?

The two chapters of Part III (this chapter and Chapter 6) have examined the movement of community gardens, their institutionalization, and their role in facilitating the development of political consciousness. The way gardeners articulate the process of their own embeddedness in the institution of community gardens and their development in them suggests that individuals who are part of the institution go through (in different ways and at different paces) a process in which they are politicized: their political awareness, consciousness, and understanding develop and grow.

What does it mean that urban residents are politicized, that they are going through a process in which their political consciousness is developing? The concept of sustainable or ecological citizenship was used to describe the actions of Los Angeles residents to revitalize the LA River and bring native animals and plants back to its banks. Ecological citizenship conceptually combines ecological understanding with civic participation:

> This form of citizenship builds on both classic ideas if citizenship, rooted in ideas about rights and obligations, but goes further to embrace a post-cosmopolitan version of citizenship. Here, rights must be weighed against obligations to those, whether close or distant, whom we have harmed historically or continue to harm on an ongoing basis. With respect to the environment, the urban ecological citizen is one whose rights include environmental justice but whose duties and obligations are defined by their ecological footprints [...]. Thus, ecological citizenship changes the geography and moves the scale of citizen action beyond

18   Ibid.

the nation-state [...] such citizenship revolves around the pursuit of ecological justice and is underpinned by the *ethics of care*.[19]

In this account, initially marginalized groups of LA activists gained influence through their actions in the pursuit of environmental justice: taking care of the natural elements that were destroyed by people, and taking care of these social groups that had to live near the destruction. The possibility of urban sustainability can materialize when ecological citizenship becomes the *mainstream*, where "many of its practices will fall within the private realm through development of a '*green consciousness*' that is apt to put individuals against dominant political-economic structures and institutions predicted on constant grow and accumulation."[20] Therefore, it is suggested, the mainstreaming of green consciousness fueled by the ethics of care and citizenship duties is a mechanism for broader social and spatial change.

The LA River project and community gardens in New York City are both examples of transforming the urban landscape in ways that afford more just distribution of resources and of the burden of externalities. Green consciousness no doubt played a crucial role in developing the vision and materializing it. But the notion of ecological citizenship, even in its mainstreamed version, does not capture all the elements that are at play in the process in which politicized urban residents practice spatial and social change. The need to deeply and broadly look into those people, urban residents, who develop a capacity to create different spaces, to relate differently to others and produce within them an alternative living environment is inspired by Lefebvre's utopian notion of *citadins* (see below). The notion of *organic residents* is proposed here to identify those people who, although they may not be defined as citadins, may nevertheless be considered a step towards these utopian urban political subjects.

Who are (or can potentially be) those organic residents? They are the truly local people, those who deal with urban conditions in actuality while the mobile elite, unaffected by the problems of the locale, provides for its needs from elsewhere. The mobile elite is in many cases the cause of fundamental local problems (for example, ghost apartment buildings) that only the truly local people are left to suffer from and deal with.[21] Despite their lesser position in the general constellation of power, the truly local people are obviously the ones that have a strong interest in changing the conditions. Therefore, where one might see a problem and weakness, another might see potential and justification for social change. Lefebvre belongs to the latter side when he advocates for urban politics of the inhabitant:

---

19   Wolch, J. 2007. Green Urban Worlds. *Annals of the Association of American Geographers* 97(2): 373–84, p. 379 (emphasis mine).

20   Ibid.

21   Bauman, Z. 2007. *Liquid Times: Living in an Age of Uncertainty*. Cambridge: Polity Press.

A radical restructuring of social, political, and economic relations, both in the city and beyond [...] reorients decision-making away from the state and toward the production of urban space [...] restructure the power relations that underlie the production of urban space [...] shifting control away from capital and the state and toward urban inhabitants.[22]

In this radical change, citizenship—the present form of enfranchisement that is separated from the right to the city—is reformed to include also rights stemming from the production of space: "enfranchisement is for those who *inhabit* the city."[23] Lefebvre called this new inhabitant-encompassing status *citadin* (citizen-inhabitant): a status that includes the right to participate and appropriate; the right to participate in decision making; and the right to physically access, occupy, and use urban space. Citadins are entitled to "the right to produce urban space that meets the needs of inhabitants"—therefore "the primary consideration in decisions that produce urban space" would be its use-value.[24]

A Lefebvrian discussion on the notion of the right to the city as it pertains to the struggle of community gardens emphasizes the conflicting rights—right to property versus right to space—epitomized in the conflict between community gardens and City Administration. The key argument against the gardens—the growing need for housing in the city that stands in conflict with the right to public space—is only the surface of a much more contentious and complex power struggle over the definition of "the public." Questions such as what the public is, what rights it has, who can claim those rights, and, in general, what the nature of city politics is are surfacing through this struggle. The gardens' struggle "is a continuing struggle over the right to the city" and the resolution of the struggle—the legal preservation of some 400 community gardens—is victorious in its potential of allowing the gardens to exist "as places where different "publics" can both exercise their right to the city and solidify that right in the landscape."[25]

With the right to the city, the city (or community gardens on a smaller scale) becomes "an *ouvre* [sic]—a work in which all its citizens participate."[26] It requires, however, active residents who maintain the struggle for the right to inhabit the city as a place that celebrates differences:

> The right to the city manifests itself as a superior form of rights: right to freedom,
> to individualization in socialization, to habitat and to inhabit. The right to the

---

22   Purcell, M. 2002. "Excavating Lefebvre: The Right to the City and its Urban Politics of the Inhabitants." *GeoJournal* 58: 99–108, pp. 101–2.

23   Ibid. (emphasis in the original).

24   Ibid., p. 103.

25   Staeheli L.A., D. Mitchell, and K. Gibson. 2002. "Conflicting Rights to the City in New York's Community Gardens." *GeoJournal* 58: 197–205, p. 198.

26   Mitchell, D. 2003. *The Right to the City: Social Justice and the Fight for Public Space*. New York: The Guilford Press, p. 17.

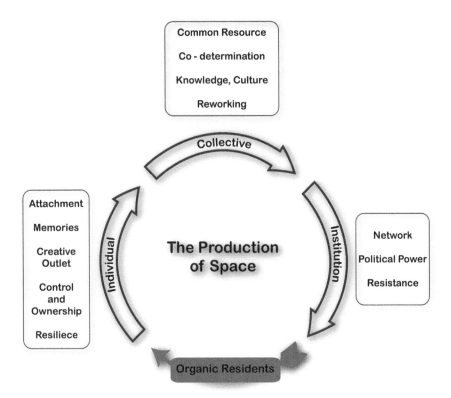

Figure 7.1   Theoretical model of levels of interaction with gardens

analysis of community gardens suggests how arguments for the right to the city, voiced by gardeners, are formulated from the ground up, from the practices over space and the experience of it.

Community gardens are sites for the development of organic residents for at least two reasons. First, community gardens are necessarily public and thus social places that, for the most part, are not experienced as private property. Most of the activities in the gardens, recreational or work and production, are of social nature and require collaboration, negotiation, and mutual learning. Second, the process of production of the space of community gardens posits every member as a participating and contributing actor (see Figure 7.1).[29] There are many different

27   Lefebvre (1996 [1968], p. 174), cited ibid. p. 18.

28   Lefebvre, H. 1995. *Introduction to Modernity*. New York: Verso, p. 116. Lefebvre uses this metaphor to explain the kind of unity that should exist between human beings and their environment.

29   Figure 7.1 presents a modular structure of the levels of interaction with the gardens. Each level is dependent on the others for its continuation. It provides insights into the process by which individual residents develop alternative consciousness, practices,

forms of involvement and even though it is possible to distinguish between active and "dormant" members, it is still important to recognize that in essence everyone is part of the production of the garden, and as such an influential social actor. The multiple opportunities for political development are embedded in many levels of the production of space.

---

and discourse that in turn fuel collective action. It also suggests the reverse processes by which those institutions that carry the agenda of the collective "reactivate" the individual participants.

Conclusion

In this book we have looked deeply into a socio-spatial phenomenon in order to understand how changes of space and of social relations are developing mutually, and in the process giving life to a different kind of urbanite. These final notes touch on what I see as the most relevant contributions of this account to broader issues of urban politics, urban nature, and urban planning and development.

Politics and the City

Contemporary political discourse is all too often dominated by a sense of the closing of alternatives, from pronouncements regarding Western capitalist social arrangements signaling "the end of history"[1] to the acknowledgment that "there is no alternative." Also contemporary geographical discourse pronouncing "the end of geography"— the outcome of time–space compression under globalization characterized by the sole domination of virtual space, annulling the social significance of all other spaces[2]—leaves us to deal with a hot, flat, and crowded world.[3]

These pronouncements are even stronger and louder when the examination focuses on the United States in general and on big global cities, such as New York, in particular. The US "is the one country in which land, from the very beginning, was treated in a manner that came closest to that dictated by purely capitalist considerations."[4] Meanwhile the global city serves as a strategic production site for the leading economic sector—corporate actors "which use the city as an 'organizational commodity'"[5]—and is rendered the terrain on which "multiplicity of globalization processes assume concrete, localized forms."[6] Extra-local interests over-determine the shape of cities and urban life; they shift attention from local public interests and well-being towards global private interests: they "brand"

1 Fukuyama, F. 1992. *The End of History and the Last Man*. New York: Harper Perennial.

2 Smith, R. 1997. "The End of Geography and Radical Politics in Baudrillard's Philosophy." *Society and Space* 15(3): 305–20.

3 Friedman, T. 2008. *Hot, Flat, and Crowded: Why We Need a Green Revolution—And How it Can Renew America*. New York: Farrar, Straus & Giroux.

4 Harvey (1982) in Logan, J. and H. Molotch. 1987. *Urban Fortunes: The Political Economy of Place*. Berkeley: University of California Press, p. 2.

5 Sassen, S. 1998. *Globalization and its Discontents*. New York: The New Press, p.xx.

6 Ibid., p.xxv.

urban centers to serve touristic and the mobile elite's purposes, and fuel processes of gentrification and the disenfranchisement of the poor and middle classes from their right to the city. All these used to be considered *faits accomplis*, realities that we can endorse or decry, but to which we inevitably need to adapt. These radical reformations of urban space (mostly molded by capital) in the last few decades were seen as apolitical concerns.

Such pronouncements not only suffer from ahistorical and apolitical fallacies, but are also deficient in their evaluation of the complexities and dialectics of space. Historically, there has always been other, multiple and contesting forms of spatial arrangements alongside the hegemonic one. There are ample examples of spontaneous, unorganized actions of using the streets, buildings, and green spaces in ways other than those prescribed—such as streets occupied by children's play,[7] homeless people's personal usage of public,[8] or the lower-class urban residents' usage of parks designed to serve the upper class.[9] Other actions that contest and redefine space are well organized and in some cases represent a social movement. Such are the historic housing movement that reshaped the housing and real estate market in New York with important policy implications that still take hold today,[10] the struggle of black people over the equal usage of public spaces such as public transportation,[11] the fight of slum residents in San Francisco against vicious slum-clearance redevelopment plans,[12] or the disability movement that changed the shape of our cities.[13] Other movements bring together a more socially eclectic group of people to protest their dissatisfaction with current use of urban space and their alternative vision of it. Examples include the "Reclaim the Streets" movement, which appropriates and transforms streets into temporary carnival or party space; and the "Critical Mass" movement, which takes over urban roads once a month using a big group of bicyclists with signs and music.[14] And recently, the economic recession followed by the election of Barack Obama as President of the United States and the discourse on a different politics, followed by unforeseen

---

7   Goodman, C. 1979. *Choosing Sides: Playground and Street Life on the Lower East Side*. New York : Schocken Books.

8   Mitchell, D. 2003. *The Right to the City: Social Justice and the Fight for Public Space*. New York: The Guilford Press.

9   Gandy, M. 2002. *Concrete and Clay: Reworking Nature in New York City*. Cambridge, MA: MIT Press.

10   Lawson, R. and M. Naison. 1986. *The Tenant Movement in New York City, 1904–1984*. New Brunswick: Rutgers University Press; Marcuse, P. 1999. "Housing Movement in the USA." *Housing, Theory and Society* 16(2): 67–86.

11   Kelly, R. 1994. *Race Rebels: Culture, Politics, and the Black Working Class*. New York: The Free Press.

12   Hartman, C. 1974. *Yerba Buena: Land Grab and Community Resistance in San Francisco*. San Francisco: Glide Publications.

13   Barnartt, S. and R. Scotch. 2001. *Disability Protests: Contentious Politics 1970–1999*. Washington, DC: Gallaudet University Press.

14   Klein, N. 1999. *No Logo: Taking Aim at the Brand Bullies*. New York: Picador.

social revolts in North Africa and the Middle East, followed by massive protests in Europe and the US all shattered the neatly fixed hegemonic picture and exposed much more of the iceberg of spatial negotiations.

This work joins others in unearthing the iceberg of spatial contestation and negotiation in order to draw attention to processes taking place on the horizon of alternatives. It was carried out because I suspected that community gardens would highlight that changes of space, its usages and representations are possible and are constantly being produced around us. Another reason was the belief that a better understanding of the intricacies of the production of alternative spaces is important for their further production. Reinforcing Lefebvre's main argument, it was shown that space is not only a product but is also a productive force. Through the struggle over space urban residents are constituted as active political actors. The gardens are both a result of social struggles and a mechanism which facilitates further struggles. They clearly confirm the power of space not only to reproduce society but also to transform it. Space is not a constant, natural, and ahistorical entity, but a dynamic and contentious result of social production. It is susceptible to intervention and disruption, and always open to change. Because spatial practices reproduce both society and the urban structure, social and spatial change is dependent on space. Acts of "intervention" in and contestation of space are not only acts *over* space but are also acts *through* space that effect and transform the reproductive relationship in society.

The case of community gardens as presented here may suggest, following Castells,[15] that the shape of resistance tends to take after the hegemonic structure. In a society defined as a network society, the movement of community gardens is organized as a relatively de-hierarchized structure with independent nodes that swarm together into action when needed. Their connection to the network is secured by personal commitment (enhanced by attachment and identification) rather than through any contractual or structural commitment. But the content of this movement, unlike its form, deviates dramatically from the hegemonic one. Residents took advantage of a crisis in the system (the economic crisis and urban disinvestment of the 1970s) to redress the needs of the have-nots and to abolish some of the injustices resulted from the uneven urban development. Finding those cracks in the system as leverage for change is how a social movement works.[16]

The case of community gardens shows that the gardens address the two sources of social injustice, putting forth a struggle for both redistribution and recognition. The gardens movement has engaged in practices aimed at reducing inequalities between groups of urbanites, most important of which is the redistribution of land. The movement struggles to integrate green open spaces into highly dense neighborhoods that for the most part lack such recreational amenities. The gardens

---

15 Castells, M. 1996. *The Rise of the Network Society: Economy, Society, and Culture*. Oxford: Blackwell.

16 Boggs, G.L. 1998. *Living for Change: An Autobiography*. Minneapolis: University of Minnesota Press.

serve as a mechanism for affording and thus redistributing a sense of ownership and control over the living environment. On a smaller scale, the gardens often help to reduce grocery expenses for residents and provide them with ample opportunities to enjoy cultural and social events free of charge.

The gardens also represent a struggle for recognition by typically celebrating differences rather than eliminating them. The most important of these is the struggle to empower the traditionally marginalized residents of the city not only to express themselves, their cultures, and their expectations of their living environment, but also to leave their mark and become significant, undismissible factors within the urban equation—to have their urban identity recognized and their right to the city asserted.

The gardens movement emphasizes two fundamental dimensions of human life, dimensions that in the last few decades were generally perceived as contradictory: the need for community and the need for self-expression. The gardens are spaces where a sense of community and self-expression (of both individuals and of groups) flourishes and inspires others. It is solidarity and cooperation that enable and enhance more significant expression of the groups and individuals within it. Furthermore, they are places where collective wisdom, among other collective resources, can be produced and shared. Much of this collective wisdom is an alternative ecological framework for the interaction between nature and the city.

Nature and the City

The integration or consolidation of nature into the city has been long debated.[17] The city has been defined as a "second nature" of stone and metal that in essence negates the "first nature" of pristine wilderness. Efforts to reconstitute nature in the city were denounced as simulations of nature and poor substitutes that can only degrade notions of nature and serve as a mask to real estate and land speculation interests in the guise of public service.[18] The analysis of community gardens sheds new light on the status of nature in the city, specifically its contribution to the construction of a different experience and notion of nature.

In actuality, the gardens are relatively small patches of green in the grand scheme of the built environment, but they afford a unique experience of nature as part of urban life. This unique experience offers a partnership in the production of nature and opportunities for caring for the natural world for its use-value rather than exploiting it for its exchange-value. More specifically, it satisfies a range of needs: for some it fulfills basic needs related to their livelihood; but it also fulfills social and psychological needs, such as the need to belong and feel connected, the need for self-

17   See for example: Williams, R. 1973. *The Country and the City*. New York: Oxford University Press; Walker, R. 2007. *The Country in the City: The Greening of the San Francisco Bay Area*. Seattle: University of Washington Press.

18   Lefebvre, H. 2003. *The Urban Revolution*. Minneapolis: University of Minnesota Press.

expression with meaningful results, and the need for ownership and control over the living environment. This kind of mutual interaction should no longer be seen as a poor substitute for first nature within second nature, but actually as potent spaces that bring nature back to the sight and thought of people.[19] Through their connection to the space of a garden residents can view the cycle of human life epitomized in the flora and fauna, dress up as flowers, bees, and other earth creatures in festivals, celebrate the seasons and harvest the fruits of the land. These experiences of nature allow gardeners to reactivate a way of being in nature that for most of us has been lost due to human domination over nature: to reconnect and be part of the web of life and to reflect on our relationship with nature and wonder at being part of it.[20] If "the goal of our highest maturity is to live and work in the world with a 'compassionate intelligence' that combines a deep identification with nature with an understanding of its processes,"[21] then community gardens afford exactly that.

In addition, it was demonstrated that nature holds another significant value for community gardens. Gardeners in community gardens use nature as a political means *and* as a politicizing means. They use ecological claims and sustainable visions to propagate their mission: to save the land of the gardens; establish more gardens; and gain more control and power over the gardens and decisions in their regard. They use the discourse on nature and sustainability as their basis for demands for equal distribution of resources; for their right for a different life in the city that is healthier, more aesthetic and more sustainable; and for their right to be recognized as equally important actors. For example, they emphasize the need for gardens as the "green lungs" of highly dense neighborhoods, and the healing power of the gardens for the community and the environment. They also define themselves as the guardians of the land *as* nature, and stress their success in taking care and maintaining the spaces for many years. In addition, through engaging with both formal and informal nature education the discourse on nature is used to further politicize the struggle of the gardens. The discourse on nature does not merely educate the public; it also politicizes it because it offers an alternative gaze and interpretation of urban life filtered through the experience of being in nature in the midst of the city. By means of education and socialization which renders nature a political claim, gardeners are able to take also the general public through a very similar process of politicization which they themselves experienced when joining their garden.

---

19   Lefebvre (1991) suggests that in urban society "the fact is that natural space will soon be lost to view [...] nature is also becoming lost to *thought*" (p. 31, emphasis in the original). "Nature is drawing away from us, to say the very least. It is becoming impossible to escape the notion that nature is being murdered by 'anti-nature'—by abstraction, by signs and images, by discourse, as also by Labour and its products" (pp. 70–71).

20   Horkheimer, M. and T. Adorno. 1998. *Dialectic of Enlightenment*. New York: Continuum.

21   Chawla, L. 2006. "Research Methods to Investigate Significant Life Experiences: Review and Recommendations." *Environmental Education Research* 12(3/4): 359–74, p. 359.

The fact that people are alienated from nature and in general have no, or very little, control over their environment is crucial to the general inability to change social reality.[22] The alternative experience of nature that the gardens provide, an experience that expands the senses—exercising the taste buds with freshly picked vegetables, providing spontaneous and formal education about people/nature relations, and generating an alternative discourse of urban sustainability—most significantly offers spaces where de-alienation and a sense of participation in one's environment can evolve reciprocally. Because of the range of alternative experiences that they offer, community gardens can be seen as spaces that cultivate a "'deep' rather than a 'shallow' form of ecological thought and practice."[23] Within the current frameworks of politics and economics, only deep ecology may engender real, sustainable change in the environment and in human relatedness to nature.

We are able to carve out alternatives in the dominant system through "prefigurations"—new forms of social and ecological arrangements that exist within and "build […] upon the transformative potentials of found configurations" can propagate a radical change in our ecological relations.[24] Community gardens are a type of prefiguration as they offer some response to many of the problematic relations with nature and present alternative relations. They offer a space where a deep reconnection with nature can happen, fostering an affection for nature while in turn strengthening a sense of competence in "managing" nature in a harmonious and sustainable manner.

As such, the gardens are spaces that facilitate "ecological integrity" which can be "maintained only by politically active citizens. Those who wait passively for public agencies, private corporations, or legislative bodies … to do the right thing are of no help at all; those institutions must always be pushed, and often opposed, by active citizens."[25] Community gardens as a movement create the community that meets, produces, and shares knowledge about ecological conditions and, at the same time, produces politically active citizens.

Local Power and Urban Planning

There are also some practical implications that may be extracted from the analysis of community gardens in New York City. The gardens can be thought of as real-life laboratories of urban planning and development as well as community organizing.

---

22   Macnaghten, P. and J. Urry, 1998. *Contested Natures*. London: Sage.

23   Biro, A. 2005. *Denaturalizing Ecological Politics: Alienation from Nature from Rousseau to the Frankfurt School and Beyond*. Toronto: University of Toronto Press, p. 13.

24   Kovel, J. 2002. *The Enemy of Nature*. London: Zed Books.

25   Tanner (1998) in Chawla, L. 2001. "Significant Life Experiences Revisited Once Again: Response to Vol. 5(4) 'Five Critical Commentaries on Significant Life Experience Research in Environmental Education.'" *Environmental Education Research* 7(4): 451–61, p. 454.

They allow for a genuine process of public participation (rather than partial or tokenistic participation when residents are consulted but do not initiate or are not part of the final decision making) that presents many desirable results.

The gardens were initiated and are operated by the residents with almost complete freedom from top-down intervention in the form of rules and regulations (complete freedom during the 1970s and almost no regulations and intervention in the 1980s and 1990s). Despite the many challenges of maintaining such an open and participatory mode of operation, there are some obvious and unique advantages to such a model. A sense of collective ownership is produced by the process of production of space that changes the interaction of residents with their living environment. Many planners and organizers would be able to argue that in this way residents become more responsible towards the environment and are more willing to volunteer time and effort in tending to it. The transference of power to the residents in public neighborhood projects can yield better results for both the community and the municipality. It may also guarantee better success with the ongoing maintenance of such participatory projects by residents without municipal intervention.

The example of parks in New York City elucidates this point. Because the city finds it hard to maintain the parks it already owns, some of the parks in the city are in such dilapidated condition that rather than serving the community they attract crime and garbage and become hazardous sites for the neighborhoods. Community gardens on the other hand are successful, for the most part, in maintaining green public sites.

Furthermore, on a more paradigmatic level there is a lot to be learned from community gardens that can be incorporated into urban planning and community organizing. The discourse, praxis, and institution of community gardens in New York City present a new form of urban conduct based on an alternative set of principles. They manifest how urban disinvestment and neglect could be addressed in creative ways to alleviate their impact on struggling populations rather than addressing it by encouraging further destruction of communities. The gardens also exhibit the rich, nuanced, and even radical knowledge that can be collectively developed if residents are given freedom of action and experience. This knowledge can be utilized to enhance any community development or community organizing project. But the greatest paradigmatic shift that is manifested in the case of community gardens is in the perceived urban power structure. Such questions as *What is the purpose of the city? Who is the city supposed to serve?* and *Whose city is it anyway?* are negotiated and in some degrees transformed in the process of production of community gardens. Community gardens challenge and reshuffle the prevailing power structure of the city. Through the production of the gardens, residents reinstitute their power as rightful participants, decision makers, and producers of urban space. They establish a situation in which "residents' needs rule." This is a productive and autopoietic situation, enabling residents to feel they have the freedom and energy to keep developing their vision of the city and acting towards its materialization. If humans are to live in urban environments, as all evidence suggests they are, community gardeners, constituting themselves as organic residents, may be able to lead us towards one way of correcting some of the imbalances characterizing social relations and relations between people and their environment.

Bibliography

Altman, I. and S. Low (eds). 2002. *Place Attachment*. New York: Plenum Press.

Apfelbaum, E. 1979. "Relation of Domination and Movement for Liberation: An Analysis of Power between Groups." In W.G Austin and S. Worchel (eds), *The Social Psychology of Intergroup Relations*. Belmont, CA: Wadsworth, pp. 188–204.

Appadurai, A. 1996. *Modernity at Large*. Minneapolis, MN: University of Minnesota Press.

Aronowitz, S. 2003. *How Class Works: Power and Social Movement*. New Haven, CT: Yale University Press.

Aronowitz, S. 2006. "Is It Time for a New Radical Party? A Meditation on Left Political Organization." *Situations* 1(2): 117–58.

Ash, A. and N. Thrift. 2002. *Cities: Reimagining the Urban*. Cambridge: Polity Press.

Barnartt, S. and R. Scotch. 2001. *Disability Protests: Contentious Politics 1970– 1999*. Washington, DC: Gallaudet University Press.

Bassett, T. 1979. "Vacant Lot Cultivation: Community Gardening in America, 1893–1978." Unpublished manuscript, Department of Geography, University of California, Berkeley.

Bauman, Z. 1998. *Globalization: The Human Consequences*. New York: Columbia University Press.

Bauman, Z. 2007. *Liquid Times: Living in an Age of Uncertainty*. Cambridge: Polity Press.

Bayat, A. 2000. "From 'Dangerous Classes' to 'Quiet Rebels': Politics of the Urban Subaltern in the Global South. *International Sociology* 15(3): 533–57.

Benner, N. and N. Theodor. 2002. "Cities and the Geographies of 'Actually Existing Neoliberalism.'" *Antipode* 34(3): 349–79.

Biro, A. 2005. *Denaturalizing Ecological Politics: Alienation from Nature from Rousseau to the Frankfurt School and Beyond*. Toronto: University of Toronto Press.

Blomley, N. 2004. *Unsettling the City: Urban Land and the Politics of Property*. New York: Routledge.

Boggs, G.L. 1998. *Living for Change: An Autobiography*. Minneapolis, MN: University of Minnesota Press.

Botkin, D.B. and C.E. Beveridge. 1997. "Cities as Environments." *Urban Ecosystems* 1(1): 3–19.

Brewer, R. 2003. *Conservancy: The Land Trust Movement in America*. Lebanon, NH: Dartmouth College Press.

198   *From the Ground Up*

Brown, B., D. Perkins, and G. Brown. 2003. "Place Attachment in a Revitalizing Neighborhood: Individual and Block Levels of Analysis." *Journal of Environmental Psychology* 23(3): 259–71.

Burgin, V. 1996. *In/different Spaces: Place and Memory in Visual Culture.* Berkeley, CA: University of California Press.

Candela, I. 2007. *Sombras de Ciudad: Arte y Transformacion Urban en Nueva York, 1970–1990.* Madrid: Alianza Editorial.

Castells, M. 1996. *The Rise of the Network Society: Economy, Society, and Culture.* Oxford: Blackwell.

Chawla, L. 2001. "Significant Life Experiences Revisited Once Again: Response to Vol. 5(4) 'Five Critical Commentaries on Significant Life Experience Research in Environmental Education.'" *Environmental Education Research* 7(4): 451–61.

Chawla, L. 2006. "Research Methods to Investigate Significant Life Experiences: Review and Recommendations." *Environmental Education Research* 12(3–4): 359–74.

Cox, R. 1999. "Civil Society at the Turn of the Millennium: Prospects for an Alternative World Order." *Review of International Studies* 25(1): 3–28.

Dagger, R. 2000. "Metropolis, Memory and Citizenship." In E. Isin (ed.), *Democracy, Citizenship and the Global City: Governance and Change in the Global Era.* London: Routledge, pp. 25–47.

Davila, A. 2004. *Barrio Dreams: Puerto Ricans, Latinos, and the Neoliberal City.* Berkeley, CA: University of California Press.

Davis, E.J. 1991. *Contested Ground: Collective Action and the Urban Neighborhood.* Ithaca, NY: Cornell University Press.

De Angelis, M. 2003. Reflections on Alternatives, Commons and Communities or Building a New World from the Bottom Up. *The Commoner* 6.

De Angelis, M. 2007 *The Beginning of History: Value Struggles and Global Capital.* London: Pluto Press.

De Certeau, M. 1988. *The Practice of Everyday Life.* Berkeley, CA: University of California Press.

DeFilippis J. 2004. *Unmaking Goliath: Community Control in the Face of Global Capital.* New York: Routledge.

Deleuze, G. 1994. *Difference and Repetition.* Trans. Paul Patton. New York: Columbia University Press.

Ehrenreich, B. 2006. *Dancing in the Streets: A History of Collective Joy.* New York: Metropolitan Books.

Eizenberg, E. 2012. Actually Existing Commons: Three Facets of Space of Community Gardens in New York City. *Antipode.* 44(3): 764–82.

Eizenberg, E. 2012. The Changing Meaning of Community Space: Two models of NGO management of community gardens in New York City. *International Journal of Urban and Regional Research* 36(1): 106–20.

Eizenberg, E. 2010. "Remembering Forgotten Landscapes: Community Gardens in New York City and the Reconstruction of Cultural Diversity." in T. Fenster and H. Yacobi (eds), *Remembering, Forgetting and City Builders.* Farnham: Ashgate, pp. 7–26.

Eizenberg, E. and T. Fenster (forthcoming). *Whose Power, Whose Autonomy? The Contested, Controlled and Neoliberalized Community Gardens.*

Etzioni, A. 1991. The Socio-Economics of Property. *Journal of Social Behavior and Personality* 6(6): 465–8.

Foucault, M. 1986. "Of Other Spaces." *Diacritics* 16(1): 22–7.

Francis, M. 1989. "Control as a Dimension of Public-Space Quality." In I. Altman and E. Zube (eds), *Public Spaces and Places.* New York: Plenum Press.

Francis, M., L. Cashdan, and L. Paxson. 1984. *Community Open Space: Greening Neighborhoods through Community Action and Land Conservation.* Washington, DC: Island Press.

Fraser, N. and A. Honneth. 2003. *Redistribution or Recognition? A Political-Philosophical Exchange.* New York: Verso.

Freire, P. 1971. "A Few Notions about the Word 'Concientization.'" *Hard Cheese* 1: 23–8.

Freire, P. 1972. *The Pedagogy of the Oppressed.* New York: Penguin.

Friedman, T. 2008. *Hot, Flat, and Crowded: Why We Need a Green Revolution—And How it Can Renew America.* New York: Farrar, Straus & Giroux.

Fukuyama, F. 1992. *The End of History and the Last Man.* New York: Harper Perennial.

Gandy, M. 2002. *Concrete and Clay: Reworking Nature in New York City.* Cambridge, MA: MIT Press.

Gibson, J. 1979. *The Ecological Approach to Visual Perception.* Hillsdale: Lawrence Erlbaum Associates.

Goldman, M. 1997. "'Customs in Common': The Epistemic World of the Commons Scholars." *Theory and Society* 26(1): 1–37.

Goodman, C. 1979. *Choosing Sides: Playground and Street Life on the Lower East Side.* New York: Schocken Books.

Hackworth, J. 2007. *The Neoliberal City: Governance, Ideology, and Development in American Urbanism.* Ithaca, NY: Cornell University Press.

Hackworth, J., and N. Smith. 2001. "The Changing State of Gentrification." *Journal of Economic and Social Geography* 92(4): 464–77.

Hancock, T. 2001. "People, Partnerships and Human Progress: Building Community Capital." *Health Promotion International* 16(3): 275–80.

Hardin, G. 1968. "The Tragedy of the Commons." *Science* 162: 1243–8.

Hardt, M. and A. Negri. 2004. *Multitude.* New York: Penguin.

Hardt, M. and A. Negri. 2009. *Commonwealth.* Cambridge, MA: Harvard University Press.

Harnik, P. 2000. *Inside City Parks.* Washington, DC: Urban Land Institute.

Hart, R. 2002. "Containing Children: Some Lessons on Planning For Play from New York City." *Environment and Urbanization* 14(2): 135–48.

Hartman, C. 1974. *Yerba Buena: Land Grab and Community Resistance in San Francisco.* San Francisco, CA: Glide Publications.

Harvey, D. 1989. *The Urban Experience*. Baltimore: Johns Hopkins University Press.

Harvey, D. 2000. *Spaces of Hope*. Berkeley, CA: University of California Press.

Harvey, D. 2003. "The City as a Body Politic." In J. Schneider and I. Susser (eds), *Wounded Cities: Destruction and Reconstruction in a Globalized World*. New York: Berg, pp. 25–44.

Harvey, D. 2006a. "Public Space and the Uses of the City." Paper presented at the conference on Refashioning Urban Spaces in Paris and New York for the 21st Century, New York University, April 29.

Harvey, D. 2006b. *Spaces of Global Capitalism: Towards a Theory of Uneven Geographical Development*. New York: Verso.

Hassell, M. 2002. *The Struggle for Eden: Community Gardens in New York City*. Westport: Bergin & Garvey.

Heft, H. 2001. *Ecological Psychology in Context: James Gibson, Roger Baker, and the Legacy of William James's Radica Empiricism*. Philadelphia, PA: Lawrence Erlbaum Associates.

Hey, R. 1998. Sense of Place in Developmental Context. *Journal of Environmental Psychology* 18: 5–29.

Highmore, B. 2002. *Everyday Life and Cultural Theory: An Introduction*. New York: Routledge.

Hiscock, R., S. Macintyre, A. Kearns, and A. Ellaway. 2003. "Residents and Residence: Factors Predicting the Health Disadvantage of Social Renters Compared to Owner-Occupiers." *Journal of Social Issues* 59(3): 527–46.

Horkheimer, M. and T. Adorno. 1998. *Dialectic of Enlightenment*. New York: Continuum.

Horton, D. 2003. "Green Distinctions: The Performance of Identity among Environmental Activists. In S. Bronislaw, H. Wallace, and C. Waterton (eds), *Nature Performed: Environment, Culture and Performance*. Oxford: Blackwell, pp. 63–78.

Isin, E. (ed.). 2000. *Democracy, Citizenship and the Global City: Governance and Change in the Global Era*. London: Routledge.

Jacobs, J. 1961. *The Death and Life of Great American Cities*. New York: Random House.

Johnson, K. 2000. "Green with Envy." City Limits (January).

Katz, C. 1998. "Whose Nature, Whose Culture? Private Production of Space and the 'Preservation' of Nature." In B. Braun and N. Castree (eds), *Remaking Reality: Nature at the Millennium*. New York: Routledge, pp. 45–62.

Katz, C. 2004. *Growing Up Global: Economic Restructuring and Children's Everyday Lives*. Minneapolis, MN: University of Minnesota Press.

Keith, M. and S. Pile (eds). 1993. *Place and the Politics of Identity*. New York: Routledge.

Kelly, R. 1994. *Race Rebels: Culture, Politics, and the Black Working Class*. New York: The Free Press.

Kingsley, J. and M. Townsend. 2006. "'Dig In' to Social Capital: Community Gardens as Mechanisms for Growing Urban Social Connectedness." *Urban Policy and Research* 24(4): 525–37.

Klein, N. 1999. *No Logo: Taking Aim at the Brand Bullies.* New York: Picador.

Kovel, J. 2002. *The Enemy of Nature.* London: Zed Books.

Laclau, E. and C. Mouffe. 2001. *Hegemony and Socialist Strategy: Towards a Radical Democratic Politics.* New York: Verso.

Latkin C. and A. Curry. 2003. "Stressful Neighborhoods and Depression: A Prospective Study of the Impact of Neighborhood Disorder." *Journal of Health and Social Behavior* 44(1): 34–44.

Lawrence, D. 1992. "Transcendence of Place: The Role of La Placeta in Valencia's Las Fallas." In I. Altman and S. Low (eds), *Place Attachment.* New York: Plenum Press, pp. 211–30.

Lawson, L. 2005. *City Bountiful: A Century of Community Gardening in America.* Berkeley, CA: University of California Press.

Lawson, R. and M. Naison. 1986. *The Tenant Movement in New York City, 1904–1984.* New Brunswick: Rutgers University Press.

Lefebvre, H. 1991. *The Production of Space.* Oxford: Blackwell.

Lefebvre, H. 1995. Introduction to Modernity. New York: Verso.

Lefebvre, H. 1996. "Part II: Right to the City." In E. Kofman and E. Lebas (eds), *Writings on Cities.* Oxford: Blackwell, pp. 63–264.

Lefebvre, H. 2002. *Critique of Everyday Life: Foundations for a Sociology of the Everyday* (vol. 2). New York: Verso.

Lefebvre, H. 2003. *The Urban Revolution.* Minneapolis, MN: University of Minnesota Press.

Light, A. 2004. "Elegy for a Garden." *Terrain.org: A Journal of the Built & Natural Environment* 15.

Linn, K. 1999. "Reclaiming the Sacred Commons." *New Village* 1: 42–9.

Lloyd, R. 2005. *Neo-Bohemia: Art and Commerce in the Postindustrial City.* New York: Routledge.

Logan, J. and H. Molotch. 1987. *Urban Fortunes: The Political Economy of Place.* Berkeley, CA: University of California Press.

Low, S. 1992. "Symbolic Ties that Bind: Place Attachment in the Plaza." In I. Altman and S. Low (eds), *Place Attachment.* New York: Plenum Press, pp. 165–84.

Low, S. and I. Altman. 1992. "Place Attachment: A Conceptual Inquiry." In I. Altman and S. Low (eds), *Place Attachment.* New York: Plenum Press, pp. 1–12.

Lulli, M. 2001. "Urban-Related Identity: Theory, Measurement, and Empirical Findings. *Journal of Environmental Psychology* 12: 285–303.

Macnaghten, P. and J. Urry, 1998. *Contested Natures.* London: Sage.

Mansfield, B. 2004. "Neoliberalism in the Oceans: 'Rationalization,' Property Rights, and the Commons Question." *Geoforum* 35(3): 313–26.

Manzo, L. 2005. "For Better or Worse: Exploring Multiple Dimensions of Place Meaning." *Journal of Environmental Psychology* 25(1): 67–86.

Marcuse, P. 1999. "Housing Movement in the USA." *Housing, Theory and Society* 16(2): 67–86.

Martinez, M. 2002. "The Struggle for the Gardens: Puerto Ricans, Redevelopment, and the Negotiation of Difference in a Changing Community." Unpublished dissertation, New York University.

Marx, K. 1959. *Economic and Philosophic Manuscripts of 1884*. Moscow: Foreign Languages Publication.

Mattila, H. 2002. "Aesthetic Justice and Urban Planning: Who Ought to Have the Right to Design Cities?" *GeoJournal* 58(2–3): 131–8.

Mayhew, M., N. Ashkanasy, T. Bramble, and J. Gardner. 2007. "A Study of the Antecedents and Consequences of Psychological Ownership in Organizational Settings." *Journal of Social Psychology* 147: 477–500.

McCann, E. 2002. "Space, Citizenship, and the Right to the City: A Brief Overview." *GeoJournal* 58: 77–9.

Mele, C. 2000. *Selling the Lower East Side: Culture, Real Estate, and Resistance in New York City*. Minneapolis, MN: University of Minnesota Press.

Merrifield, A. 2005. *Henri Lefebvre: A Critical Introduction*. New York: Routledge.

Mitchell, D. 1995. "The End of Public Space? People's Park, Definitions of Public and Democracy." *Annals of the Association of American Geographers* 85: 108–33.

Mitchell, D. 2003. *The Right to the City: Social Justice and the Fight for Public Space*. New York: The Guilford Press.

Monbiot, G. 1994. "The Tragedy of Enclosure." *Scientific American* 27(1): 159.

Morrison, K. 2006. *Marx, Durkheim, Weber: Formations of Modern Social Thought*. Thousand Oaks: Sage.

Nasar J. and B. Fisher 1993. "'Hot Spots' of Fear and Crime: A Multi-Method Investigation." *Journal of Environmental Psychology* 13: 187–206.

Parker, S., T. Wall, and P. Jackson. 1997. "That's Not My Job": Developing Flexible Employee Work Orientations. *Academy of Management Journal* 40: 899–929.

Peck, J. and A. Tickel. 2002. "Neoliberalizing Space." *Antipode* 34(3): 380–404.

Pierce, J., M. O'Driscoll, and A. Coghlan. 2004. "Work Environment Structure and Psychological Ownership: The Mediating Effect of Control." *Journal of Social Psychology* 144: 507–34.

Pile, S. 1996. *The Body and the City: Psychoanalysis, Space and Subjectivity*. New York: Routledge.

Polanyi, K. 2001. *The Great Transformation: The Political and Economic Origins of Our Time*. Boston, MA: Beacon Press.

Proshansky, H., A. Fabian, and R. Kaminoff. 1983. "Place-Identity: Physical World Socialization of the Self." *Journal of Environmental Psychology* 3: 57–83.

Pudup, M.B. 2008. It Takes a Garden: Cultivating Citizen-Subjects in Organized Garden Projects. *Geoforum* 39: 1228–40.

Purcell, M. 2002. "Excavating Lefebvre: The Right to the City and its Urban Politics of the Inhabitants." *GeoJournal* 58: 99–108.

Quastel, N. 2009. "Political Ecology of Gentrification." *Urban Geography* 30(7): 694–725.

Rose, N. 1998. *Inventing Our Selves: Psychology, Power, and Personhood.* Cambridge: Cambridge University Press.

Rose, N. 1999. *Powers of Freedom: Reframing Political Thought.* Cambridge: Cambridge University Press.

Rosol, M. 2012. "Community Gardens as Neoliberal Strategy? Green Space Production in Berlin." *Antipode* 44(1): 239–57.

Saegert, S. 2006. "Building Civic Capacity in Urban Neighborhoods: An Empirically Grounded Anatomy." *Journal of Urban Affairs* 28(3): 275–94.

Saegert, S. and L. Benitez. 2005. "Limited Equity Housing Cooperatives: Defining a Niche in the Low Income Housing Market." *Journal of Planning Literature* 19(4): 427–39.

Saldivar-Tanaka, L. and M.E. Krasny. 2004. "Culturing Community Development, Neighborhood Open Space, and Civic Agriculture: The Case of Latino Community Gardens in New York City." *Agriculture and Human Values* 21: 399–412.

Sandercock, L. 1998. *Towards Cosmopolis.* London: Wiley.

Sassen, S. 1998. *Globalization and its Discontents.* New York: The New Press.

Schmelzkopf, K. 1995. "Urban Community Gardens as a Contested Space." *Geographical Review* 85(3): 364–81.

Schmelzkopf, K. 2002. "Incommensurability, Land Use, and the Right to Space; Community Gardens in New York City." *Urban Geography* 23(4): 323–43.

Shepard, B. 2005. "Play, Creativity, and the New Community Organizing." *Journal of Progressive Human Services* 16(2): 47–69.

Shepard, B. and R. Hayduk. 2002. *From ACT UP to the WTO: Urban Protest and Community Building in the Era of Globalization.* New York: Verso.

Site, W. 2003. *Remaking New York: Primitive Globalization and the Politics of Urban Community.* Minneapolis, MN: University of Minnesota Press.

Smith, N. 2002. "New Globalism, New Urbanism: Gentrification as Global Urban Strategy." *Antipode* 34(3): 427–50.

Smith, R. 1997. "The End of Geography and Radical Politics in Baudrillard's Philosophy." *Society and Space* 15(3): 305–20.

Soja, E. 1989. *Postmodern Geographies: The Reassertion of Space in Critical Social Theory.* New York: Verso.

Sorkin, M. 1992. *Variations on a Theme Park: The New American City and the End of Public Space.* New York: Hill and Wang.

Staeheli, L.A., D. Mitchell, and K. Gibson. 2002. "Conflicting Rights to the City in New York's Community Gardens." *GeoJournal* 58: 197–205.

Tajbakhsh, K. 2001. *The Promise of the City: Space, Identity, and Politics in the Contemporary Social Thought.* Berkeley, CA: University of California Press.

Thabit, W. 2003. *How East New York Became a Ghetto*. New York: New York University Press.

Vaske, J. and K. Kobrin. 2001. "Place Attachment and Environmentally Responsible Behavior." *Journal of Environmental Education* 32(4): 16–21.

Voicu, I. and V. Been. 2008. "The Effect of Community Gardens on Neighboring Property Values." *Real Estate Economics* 36(2): 241–83.

Vygotsky, L.S. 1978. *Mind in Society: The Development of Higher Psychological Processes*. Cambridge, MA: Harvard University Press.

Walker, R. 2007. *The Country in the City: The Greening of the San Francisco Bay Area*. Seattle: University of Washington Press.

Williams, R. 1973. *The Country and the City*. New York: Oxford University Press.

Wolch, J. 2007. "Green Urban Worlds." *Annals of the Association of American Geographers* 97(2): 373–84.

Zukin, S. 2010. Naked City: *The Death and Life of Authentic Urban Places.* Oxford: Oxford University Press.

Index